Structural Plasticity

W.F. Chen H. Zhang

Structural Plasticity

Theory, Problems, and CAE Software

With 102 Illustrations and a Diskette

Springer-Verlag
New York Berlin Heidelberg London
Paris Tokyo Hong Kong Barcelona

W.F. Chen
Purdue University
West Lafayette, IN
USA

H. Zhang
Purdue University
West Lafayette, IN
USA

Library of Congress Cataloging-in-Publication Data
Chen, Wai-Fah, 1936–
Structural plasticity : theory, problems, and CAE software / W.F. Chen, H. Zhang.
P. cm.
Includes index.
ISBN 0-387-96789-3 (New York). — ISBN 3-540-96789-3 (Berlin)
1. Plasticity. 2. Structural design. 3. Structural analysis (Engineering)—Data processing. I. Zhang, H. II. Title.
TA418.14.C49 1988 Suppl.
620.1'1233—dc20 90-19240
CIP

Printed on acid-free paper.

Camera-ready copy provided by the authors using *Troff*.
Printed and bound by Edwards Brothers, Inc., Ann Arbor, Michigan.
Printed in the United States of America.

9 8 7 6 5 4 3 2 1

ISBN 0-387-96789-3 Springer-Verlag New York Berlin Heidelberg
ISBN 3-540-96789-3 Springer-Verlag Berlin Heidelberg New York

Preface

This book is designed for use as a supplement to the textbook on "PLASTICITY FOR STRUCTURAL ENGINEERS" by W. F. Chen and D. J. Han (Springer-Verlag, 1988). The purpose is to help students and structural engineers to learn and to practice how to solve typical engineering plasticity problems in general and, more importantly, how to use computers to solve plasticity problems in structural engineering in particular. To this end, specific numerical algorithms in the computer software implementation of the theory together with actual code development are given.

The book is divided into six chapters in a logical sequence. First, the relevant plasticity theory involved is briefly reviewed at the beginning of each chapter. A number of solved and supplementary problems with answers are then provided. The solved problems serve to illustrate and amplify the theory, to bring into a sharp focus those concepts or procedures without which the students continuously feel themselves on an uncomfortable ground, and to provide the necessary exercise or practice so vital to an effective learning. The supplementary problems can be used for further exercises.

Two Computer-Aided-Education (CAE) programs for structural plasticity, PLASTIC1 and PLASTIC_ZONE, are introduced in Chapter 2 and Section 5.9 respectively. The major part of these two programs is written in C language, but the other two languages YACC (Yet Another Compiler-Compiler) and LEX (A Lexical Analyzer Generator) developed at Bell Laboratories are also used in the program PLASTIC1 for interpreting commands. The compilers for YACC and LEX are available in most UNIX systems. In the original planning of this book, more CAE programs are to be included. However, these programs are not readily available for inclusion at this time. In the next edition, more CAE programs for structural plasticity will be provided. At present, they are under active development at Purdue University.

September, 1990
West Lafayette, Indiana

W.F. Chen
H. Zhang

Contents

Notation

Stresses and Strains

σ_{ij}	stress tensor
$\sigma_1, \sigma_2, \sigma_3$	principal components of stress tensor, tensile stress positive
s_{ij}	deviatoric stress tensor
s_1, s_2, s_3	principal components of deviatoric stress tensor
σ	normal stress
τ	shear stress
ε_{ij}	strain tensor
$\varepsilon_1, \varepsilon_2, \varepsilon_3$	principal components of strain tensor, tensile strain positive
ε_{ij}^e	elastic strain tensor
ε_{ij}^p	plastic strain tensor
ε	normal strain
γ	engineering shear strain
ε_v	volumetric strain
e_{ij}	deviatoric strain tensor
e_1, e_2, e_3	principal components of deviatoric strain tensor
$f(\sigma_{ij}, \varepsilon_{ij}^p, \kappa)$	yield function or loading function
$g(\sigma_{ij}, \varepsilon_{ij}^p, \kappa)$	plastic potential function
σ_e	effective stress
α_{ij}	coordinate of the center of loading surfaces
$\varepsilon_p = \int (d\varepsilon_{ij}^p \, d\varepsilon_{ij}^p)^{1/2}$	equivalent or effective plastic strain
$W_p = \int \sigma_{ij} \, d\varepsilon_{ij}^p$	plastic work

Invariants

$I_1 = \sigma_{ii}$	first invariant of stress tensor
$J_2 = \frac{1}{2} s_{ij} s_{ji}$	second invariant of deviatoric stress tensor

$J_3 = \frac{1}{3} s_{ij} s_{jk} s_{ki}$	third invariant of deviatoric stress tensor
$\cos 3\theta = \frac{3\sqrt{3}}{2} \frac{J_3}{J_2^{3/2}}$	where θ is the angle of similarity defined in Fig. 3.2
$I_1' = \varepsilon_{ii}$	first invariant of strain tensor
$J_2' = \frac{1}{2} e_{ij} e_{ji}$	second invariant of deviatoric strain tensor
$\rho = \sqrt{2 J_2}$	deviatoric length defined in Fig. 3.1
$\xi = \frac{1}{\sqrt{3}} I_1$	hydrostatic length defined in Fig. 3.1

Material Parameters

C_{ijkl}	elastic stiffness tensor
C^{p}_{ijkl}	plastic stiffness tensor
C^{ep}_{ijkl}	elastic-plastic stiffness tensor
K	bulk modulus
G	shear modulus
E	Young modulus
ν	Poisson ratio
E_p	plastic modulus in uniaxial stress condition
E_t	tangential modulus in uniaxial stress condition
H_p	slope of effective stress-effective plastic strain curve
σ_0	initial yield stress in uniaxial tension
τ_0	initial yield stress in pure shear
σ_t	uniaxial tensile yield stress ($\sigma_t > 0$)
σ_c	uniaxial compressive yield stress ($\sigma_c = m\,\sigma_t > 0$)
k	pure shear yield stress
c, ϕ	cohesion and friction angle in Mohr-Coulomb criterion
$m = \sigma_c / \sigma_t$	parameter in Mohr-Coulomb criterion
α, k	strength parameter in Drucker-Prager criterion
κ	hardening parameter or plastic internal variable

Chapter 1

One-Dimensional Stress-Strain Analysis

Uniaxial stress problems deal with structures having only one non-zero principal stress component. This non-zero component can be either positive (uniaxial tension) or negative (uniaxial compression). It is generally necessary to study one-dimensional problems in structural plasticity in order to understand the elastic-plastic behavior of materials in structures under complex general stress conditions.

1.1 Uniaxial Behaviors

1.1.1 Monotonic Loading

The typical stress-strain behavior of metals in simple tension is shown in Fig. 1.1. Up to the stress level σ_{y0} at point A_0, the response is *linearly elastic*, and unloading follows the same initial loading path. The range O-A is called *elastic range*, and loading in this range is called *elastic loading*. The behavior of the material in this range is *load path independent*, and stress determines strain uniquely, and vice verse.

The stress state at point A_0 is called *elastic limit*, and for practical purpose it is also known as *initial yield point*. For stress and strain states beyond point A_0 and into the region A_0–C, the material becomes plastic and behaves *irreversibly*. In this plastic range, the stiffness or the slope of the stress-strain curve decreases progressively, and eventually the material fails at point C. This is called *plastic loading*.

1.1.2 Unloading and Reloading

Unloading from a point in the plastic range A_0–C, say, point A_1, results in a proportionally decreasing stress and strain along the line $\overline{A_1O_1}$ parallel to the initial elastic loading path $\overline{O_0A_0}$. A complete unloading from point A_1 to zero at point O_1 leaves a *permanent strain* or *plastic strain* ε_1^p. Only a part of the total strain, ε_1, at A_1 is recovered upon unloading, called *elastic strain* ε_1^e. Subsequent reloading from point O_1

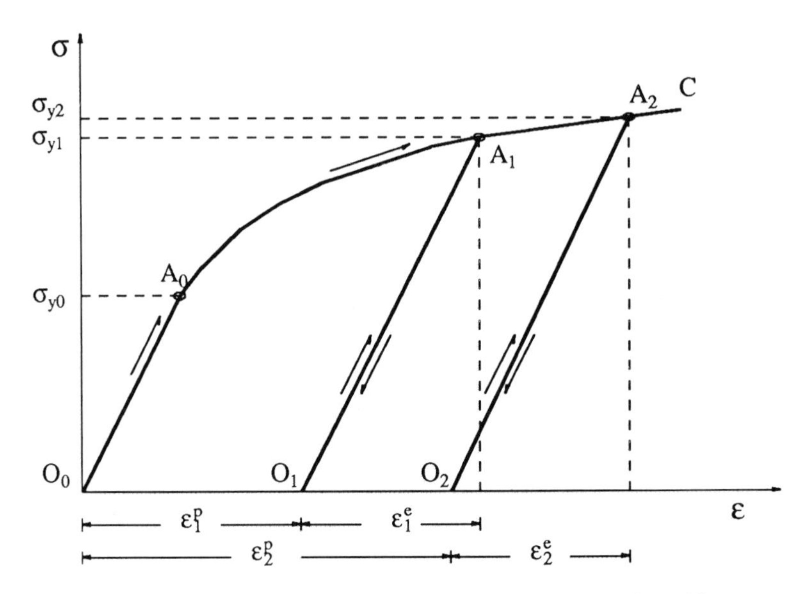

Figure 1.1. Typical elastic-plastic stress-strain relationship

proceeds along the same line $\overline{O_1A_1}$ up to point A_1, and further plastic loading follows the original nonlinear curve A_1–C, as if unloading and reloading had never taken place.

1.1.3 Reversed Loading

For metals, the stress-strain curve in a simple compression test is usually identical with that in a simple tension test. On the other hand, if the material is first loaded in tension up to the plastic state, and then unloaded and reloaded in compression, the stress-strain curve of the material in compression can be significantly different from the curve that would be obtained directly from a simple compression test without prior tension loading. This is illustrated in Fig. 1.2. The new yield point in compression at point B corresponds to stress σ_B. σ_B is smaller than σ_{y0} and is much smaller than the *subsequent yield stress* at point A. This phenomenon is called *Bauschinger Effect.*

It is apparent that the stress-strain behavior in the plastic range is *load path dependent*, and there is no one-to-one coordination between stress and strain. Generally speaking, the strain state depends not only on current stress state, but also on the entire prior loading history, i.e., *stress history* as well as *deformation history.* This is exemplified in Fig. 1.1 in which the zero stress states correspond to three different deformation

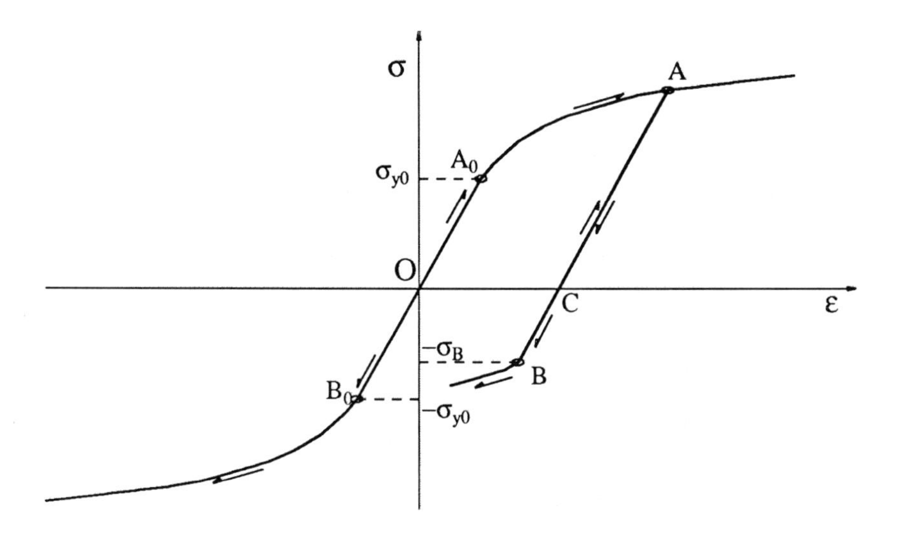

Figure 1.2. Bauschinger effect on reversed loading

states O_0, O_1, and O_2.

1.2 Basic Relations

Since there is no one-to-one coordination between stress and strain in the plastic state, it is not possible to express the stress-strain relationship in terms of total stress and total strain. Thus, for elastic-plastic materials, only a unique incremental relationship between stress and strain increments can be written and expressed in terms of the stress and deformation history. This is illustrated in Fig. 1.3.

In Fig.1.3, the strain increment $d\varepsilon$ is decomposed into two parts: the elastic strain increment $d\varepsilon^e$ and the plastic strain increment $d\varepsilon^p$, and the general incremental relationship can then be written as

$$d\varepsilon = d\varepsilon^e + d\varepsilon^p \tag{1.1}$$

$$d\sigma = E_t\, d\varepsilon = E\, d\varepsilon^e = E_p\, d\varepsilon^p \tag{1.2}$$

where $d\sigma$ is the corresponding stress increment, E the Young's modulus, and E_t and E_p the *tangential* modulus and *plastic* modulus respectively. These moduli are functions of deformation history. For a given deformation history, they may be derived from an experimental stress-

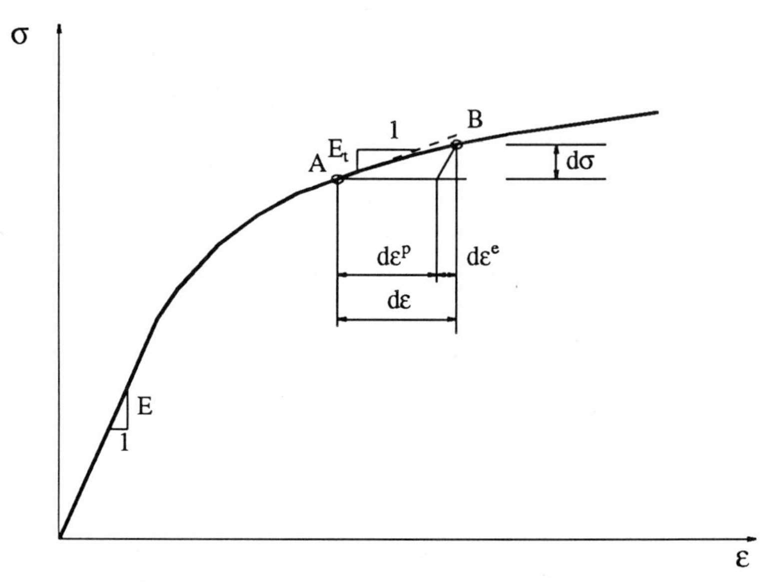

Figure 1.3. Stress and strain increments

strain curve under a monotonic loading condition with an assumed hardening rule. This will be discussed in the forthcoming.

Note that

$$E_t = \frac{d\sigma}{d\varepsilon}, \quad E_p = \frac{d\sigma}{d\varepsilon^p} \tag{1.3}$$

from which, we obtain the relation

$$\frac{1}{E_t} = \frac{1}{E} + \frac{1}{E_p} \tag{1.4}$$

or

$$E_t = \frac{E\,E_p}{E + E_p}, \quad E_p = \frac{E\,E_t}{E - E_t} \tag{1.5}$$

For a given elastic-plastic state, a stress increment or a strain increment may cause plastic loading or elastic unloading. In the case of plastic loading, new plastic deformation accumulates. In the case of elastic unloading, no new plastic deformation occurs, and the relationship between stress and strain increments follows the elastic rule

$$d\sigma = E\,d\varepsilon \tag{1.6}$$

The condition determining whether a stress increment or a strain increment causes plastic loading or elastic unloading is called *loading criterion.* For uniaxial stress condition, it has the simple form

$$\sigma = \sigma_y \quad \text{and} \quad \sigma\, d\sigma > 0 \tag{1.7}$$

where σ_y is the maximum stress value ever reached in the stress history. Alternatively, in terms of strain and strain increment, the loading criterion can be expressed as

$$\varepsilon = \varepsilon_y \quad \text{and} \quad \varepsilon\, d\varepsilon > 0 \tag{1.8}$$

where ε_y is the maximum strain value ever reached in the strain history. Note that there are two maximum stress values: one for tension loading and the other for compression loading. They are not the same in general and depend on the hardening rule. Similar statement can be made for the maximum strain values. Equations (1.1), (1.2), (1.7) and (1.8) are valid for very small stress and strain increments. If the increments are finite, Equations (1.1) and (1.2) must be integrated.

To obtain the solution for an elastic-plastic problem, the actual elastic-plastic material behavior must be idealized. For one-dimensional problems, the elastic-plastic behavior can be represented by idealized stress-strain relations under a monotonic loading together with an assumed hardening rule. In the following two sections, idealized stress-strain relationships and hardening rules will be presented.

1.3 Idealized Stress-Strain Relationships

There are many stress-strain models available for the elastic-plastic behavior under a monotonic loading, and can all be represented by either simple analytical expressions or sets of discrete stress-strain points. Some frequently used analytical models are listed in the following for the stress-strain relationships under the monotonic tension state.

1.3.1 Elastic-Perfectly Plastic Model

In this model the work hardening effect is neglected. Plastic flow begins when stress reaches the yield stress σ_{y0}.

$$\varepsilon = \frac{\sigma}{E} \qquad \text{for } \sigma < \sigma_{y0} \tag{1.9a}$$

$$\varepsilon = \frac{\sigma}{E} + \lambda \qquad \text{for } \sigma = \sigma_{y0} \tag{1.9b}$$

where λ is a positive scaler. Note that when stress $\sigma = \sigma_{y0}$ is reached, the corresponding strain is undefined. The magnitude of strain can be determined only by the surrounding elastic material restricting the plastic flow in a partially elastic and partially plastic structure.

1.3.2 Elastic-Linearly Hardening Model

In this model the tangential modulus is assumed to be constant, and the stress-strain relation is approximated by two straight lines.

$$\varepsilon = \frac{\sigma}{E} \qquad \text{for } \sigma \le \sigma_{y0} \tag{1.10a}$$

$$\varepsilon = \frac{\sigma}{E} + \frac{1}{E_t}(\sigma - \sigma_{y0}) \qquad \text{for } \sigma > \sigma_{y0} \tag{1.10b}$$

1.3.3 Elastic-Exponential Hardening Model

In this model, the stress-strain relation is divided into two parts: a linear expression for the elastic region and a power expression for the elastic-plastic region.

$$\sigma = E\,\varepsilon \qquad \text{for } \sigma \le \sigma_{y0} \tag{1.11a}$$

$$\sigma = k\,\varepsilon^n \qquad \text{for } \sigma > \sigma_{y0} \tag{1.11b}$$

where k and n are material constants to be determined by a curve-fitting with the experimental curve.

1.3.4 Ramberg-Osgood Model

In this model, a nonlinear smooth expression is employed to represent the entire stress-strain relation

$$\varepsilon = \frac{\sigma}{E} + a\left(\frac{\sigma}{b}\right)^n \tag{1.12}$$

where a, b and n are material constants to be determined by a curve-fitting with the experimental curve.

1.4 Hardening Rules

1.4.1 The Hardening Parameter

A hardening rule describes a specific relationship between the subsequent yield stress σ_y of a material and the plastic deformation accumulated during prior plastic loadings. To begin with, we need first define a state variable to quantitatively represent plastic deformation. Such a variable, called *hardening parameter* or *plastic internal variable*, is often represented by the symbol κ. The subsequent yield stresses σ_y, tangential modulus E_t, and plastic modulus E_p can all be expressed as functions of the hardening parameter κ.

The hardening parameter κ can be defined in several ways. The three frequently used definitions are listed below.

$$\kappa = \varepsilon_p = \int \sqrt{d\varepsilon^p \, d\varepsilon^p} \quad \text{equivalent plastic strain} \quad (1.13)$$

$$\kappa = W_p = \int \sigma \, d\varepsilon^p \quad \text{plastic work} \quad (1.14)$$

$$\kappa = \varepsilon^p = \int d\varepsilon^p \quad \text{plastic strain} \quad (1.15)$$

These three definitions represent in different ways the same physical quantity, the plastic deformation. The actual functional representation of a hardening rule varies according to different definitions of the hardening parameter.

Note that the plastic strain ε^p is not a good state variable to represent the history of plastic deformation because ε^p is reversible and cannot be accumulated. For example, a bar can be pulled to the plastic range with ε^p, and then compressed back to its original length. The actual total plastic strain accumulated under this loading history is $2\varepsilon^p$, while the definition of Eq.(1.15) indicates a zero plastic deformation history for this loading path.

1.4.2 Idealized Hardening Rules

A hardening rule expresses the relationship of the subsequent yield stress, tangential modulus or plastic modulus with the hardening parameter. The plastic deformation accumulated in a tension loading affects not only the yield stress in tension, but also its subsequent yield stress in compression, as reflected by the Bauschinger effect. This is also true for the plastic deformation accumulated in a compression loading. Different modelings of such effects result in different hardening rules. However, if the loading direction is not changed, the hardening behavior represented by different hardening rules should be identical. Three types of idealized hardening rules are described in the following.

1. Isotropic hardening rule: The isotropic hardening rule states that the progressively increasing yield stresses under both tension and compression loadings are always the same. Thus, this hardening rule is expressed as

$$|\sigma| = |\sigma(\kappa)| \tag{1.16}$$

2. Kinematic hardening rule: The kinematic hardening rule states that the difference between the yield stresses under tension loading and under compression loading remains a constant. Let σ_y^t denotes the yield stress under a tension loading, while σ_y^c denotes the yield stress under a compression loading, this hardening rule is expressed as

$$\sigma_y^t(\kappa) - \sigma_y^c(\kappa) = 2\,\sigma_{y0} \tag{1.17a}$$

where σ_{y0} is the initial yield stress, and the initial yield stresses in tension and in compression are assumed equal. The kinematic hardening rule can also be expressed conveniently in the form as

$$|\sigma - c(\kappa)| = \sigma_{y0} \tag{1.17b}$$

where the function $c(\kappa)$ represents the track of the elastic center and satisfies $c(0) = 0$.

3. Independent hardening rule: The independent hardening rule states that the subsequent yield stresses under a tension loading and under a compression loading are independent to each other. The plastic deformation accumulated in a tension loading history only affects the tension yield stress, while the plastic deformation accumulated under a compression loading only affects the compression yield stress. Thus, two hardening parameters, κ^t and κ^c, are used to represent the two different types of plastic deformations respectively. The independent hardening rule is then expressed as

$$\sigma = \sigma_y^t(\kappa^t) \qquad \text{for } \sigma > 0 \tag{1.18a}$$

$$\sigma - \sigma_y^c(\kappa^c) \qquad \text{for } \sigma < 0 \tag{1.18b}$$

There can be also a mixed hardening rule that combines the behavior of both isotropic and kinematic hardening rules. A mixed hardening rule will be discussed in the next chapter where the rule will be implemented in a software to solve one-dimensional problems numerically.

1.5 Stress-Strain Response Problems

Prob. 1.1 The $\sigma - \varepsilon$ response in simple tension is approximated by the following linear work-hardening model

$$\sigma = E\,\varepsilon, \qquad 0 \le \varepsilon \le \varepsilon_0$$

$$\sigma = \sigma_0 + m\,\varepsilon^p, \qquad \varepsilon_0 \le \varepsilon, \qquad \varepsilon_0 = \sigma_0 / E$$

where $m = E/9 = \text{const}$. An element of the material undergoes the following uniaxial stress history: $\sigma = 0 \to \sigma_1 \to 0 \to -\sigma_1 \to 0$, where $\sigma_1 = (1.0 + \alpha)\,\sigma_0$, $(0 \le \alpha < 1)$. Determine the strain history corresponding to the given stress history, and the plastic strain, plastic work and plastic equivalent strain at the end of the loading and unloading history. Sketch the stress vs. strain curve. Assume isotropic hardening rule.

Solution: For a given stress, strain, and plastic deformation state, the response of an elastic-plastic material to a given stress increment depends on whether the stress increment constitutes an elastic loading, plastic loading, or an unloading. In the case of plastic loading, the strain increment can be expressed as

$$d\varepsilon = d\sigma / E_t(\kappa)$$

where E_t is the tangential modulus, and κ is the plastic internal variable. In the case of elastic loading or unloading,

$$d\varepsilon = d\sigma / E$$

Therefore, for a given stress increment, the loading state must be checked first.

Plastic deformation is produced only in a plastic loading process. For a given stress increment constituting a plastic loading, we have

$$d\varepsilon^p = d\sigma / E_p$$

$$d\varepsilon_p = |d\sigma| / E_p$$

and

$$dW_p = \sigma\, d\varepsilon^p = \sigma\, d\sigma / E_p$$

where E_p is the plastic modulus.

For a given finite stress increment $\Delta\sigma$, the above equations need to be integrated in general to obtain a solution. However, for the present problem, the plastic modulus and tangential modulus are constants, $E_p = m = E/9$, and

$$E_t = \frac{EE_p}{E + E_p} = \frac{E}{10}$$

The differential increment can be replaced with a finite increment, except for the expression of plastic work, W_p. The increment of plastic work corresponding to a finite stress increment can be expressed as, at a given stress state

$$\Delta W_p = \frac{1}{E_p} \int_{\sigma}^{\sigma+\Delta\sigma} \sigma \, d\sigma = \frac{1}{2E_p}[\Delta\sigma(2\sigma + \Delta\sigma)]$$

The complete stress-strain history is plotted as solid curve in Fig. S1.1 for $\alpha = 0.4$. The material is initially elastic until it reaches the initial yield point at Point A as shown in Fig. S1.1, where $\sigma_A = \sigma_0$, $\varepsilon_A = \varepsilon_0 = \sigma_0 / E$. From the initial yield point to Point B where $\sigma_B = \sigma_0 + \Delta\sigma = \sigma_1$, the material is in the plastic range. The relationship $\Delta\sigma = E_t \Delta\varepsilon$ should be used. Thus, we obtain at $\sigma_B = \sigma_1 = (1 + \alpha)\, \sigma_0$, Point B,

$$\varepsilon_B = \varepsilon_0 + \Delta\varepsilon = \varepsilon_0 + (\sigma_1 - \sigma_0)/E_t = (1 + 10\alpha)\, \varepsilon_0$$

and

$$\varepsilon_B^p = \Delta\sigma / E_p = (\sigma_1 - \sigma_0)/E_p = 9\alpha\, \varepsilon_0, \qquad \varepsilon_{pB} = 9\alpha\, \varepsilon_0$$

$$W_{pB} = W_{pA} + \Delta W = 9(\frac{1}{2}\alpha^2 + \alpha)\, \sigma_0 \varepsilon_0$$

From $\sigma = \sigma_1$ to $\sigma = 0$, the material unloads. We obtain at $\sigma = 0$, Point C,

$$\varepsilon_C = \varepsilon_B - \sigma_B / E = (1 + 10\alpha)\, \varepsilon_0 - \sigma_B / E = 9\alpha\varepsilon_0$$

No plastic deformation is produced in this process. From $\sigma = 0$ to $\sigma = -\sigma_B$, the material is in a reversed loading. According to the isotropic hardening rule, the material yields in compression at Point D_1 where $\sigma_{D_1} = -\sigma_B$, the end of this loading path. Therefore, no plastic deformation occurs in this path. We obtain at Point D_1,

$$\varepsilon_{D_1} = \varepsilon_C - \sigma_B / E = 9\alpha\, \varepsilon_0 - \sigma_1 / E = (8\alpha - 1)\, \varepsilon_0$$

From $\sigma = -\sigma_1$ to $\sigma = 0$, the material unloads, and at Point C where $\sigma_C = 0$, we obtain

$$\varepsilon_C = \varepsilon_{D_1} + \sigma_B / E = (8\alpha - 1)\, \varepsilon_0 + \sigma_1 / E = 9\alpha\, \varepsilon_0$$

The complete stress-strain history is given below.

$$[\varepsilon\,,\, \sigma] = [0\,,\, 0]_O \rightarrow [\varepsilon_0\,,\, \sigma_0]_A \rightarrow [(1 + 10\alpha)\, \varepsilon_0\,,\, (1 + \alpha)\, \sigma_0]_B \rightarrow$$
$$[9\alpha\varepsilon_0\,,\, 0]_C \rightarrow [(8\alpha - 1)\, \varepsilon_0\,,\, -(1 + \alpha)\, \sigma_0]_{D_1} \rightarrow [9\alpha\, \varepsilon_0\,,\, 0]_C$$

At the end of this stress history, the plastic strain, plastic equivalent strain, and plastic work are, respectively

$$\varepsilon^{Pc} = 9\alpha\varepsilon_0\,, \quad \varepsilon_{pC} = 9\alpha\varepsilon_0\,, \quad W_{pC} = 9(\frac{1}{2}\alpha^2 + \alpha)\, \sigma_0\varepsilon_0$$

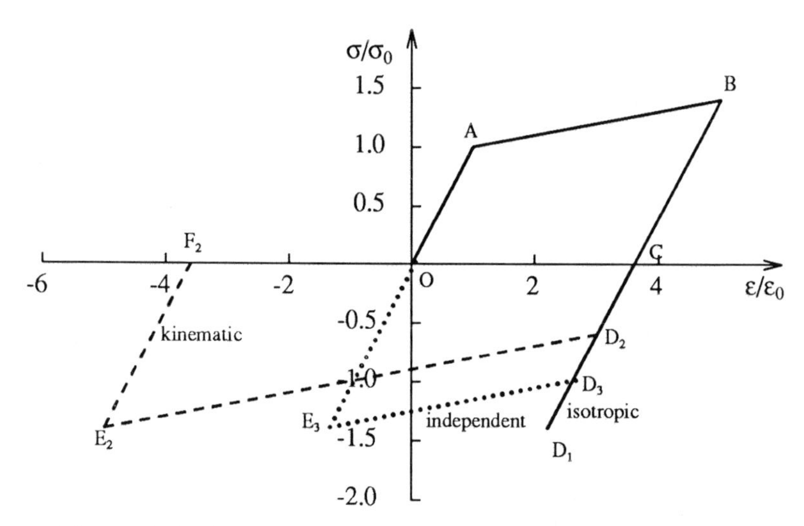

Figure S1.1. Stress vs. strain relations for Prob. 1.1 to Prob. 1.3

Prob. 1.2 The same stress-strain response problem as Prob. 1.1, except that the material is assumed to follow kinematic hardening rule.

Hint: The material yields in the reversed compression loading at $\sigma = -(2\sigma_0 - \sigma_B)$.

Answer: The complete stress-strain history is

$$[\varepsilon\ ,\ \sigma] = [0\ ,\ 0]_O \rightarrow [\varepsilon_0\ ,\ \sigma_0]_A \rightarrow [(1 + 10\alpha)\ \varepsilon_0\ ,\ (1 + \alpha)\ \sigma_0]_B \rightarrow$$

$$[9\alpha\varepsilon_0\ ,\ 0]_C \rightarrow [(10\alpha - 1)\ \varepsilon_0\ ,\ -(1 - \alpha)\ \sigma_0]_{D_2} \rightarrow$$

$$[-(1 + 10\alpha)\ \varepsilon_0\ ,\ -(1 + \alpha)\ \sigma_0]_{E_2} \rightarrow [-9\alpha\ \varepsilon_0\ ,\ 0]_{F_2}$$

and the stress vs. strain curve is plotted in Fig. S1.1. At the end of this stress history, the plastic strain, plastic equivalent strain, and plastic work are, respectively

$$\varepsilon_F^p = -9\alpha\varepsilon_0\ , \quad \varepsilon_{pF} = 27\alpha\varepsilon_0\ , \quad W_{pF} = 9(\frac{1}{2}\alpha^2 + 3\alpha)\ \sigma_0\varepsilon_0$$

Prob. 1.3 The same stress-strain response problem as Prob. 1.1, except that the material is assumed to follow independent hardening rule.

Hint: The material yields in the reversed compression loading at $\sigma = -\sigma_0$.

Answer: The complete stress-strain history is

$$[\varepsilon , \sigma] = [0 , 0]_O \rightarrow [\varepsilon_0 , \sigma_0]_A \rightarrow [(1 + 10\alpha)\, \varepsilon_0 , (1 + \alpha)\, \sigma_0]_B \rightarrow$$
$$[9\alpha\varepsilon_0 , 0]_C \rightarrow [(9\alpha - 1)\, \varepsilon_0 , -\sigma_0]_{D_3} \rightarrow$$
$$[-(1 + \alpha)\, \varepsilon_0 , -(1 + \alpha)\, \sigma_0]_{E_3} \rightarrow [0 , 0]_O$$

and the stress vs. strain curve is plotted as dotted curve in Fig. S1.1. At the end of this stress history, the plastic strain, plastic equivalent strain, and plastic work are, respectively

$$\varepsilon^p_O = 0 , \quad \varepsilon_{pO} = 18\alpha\varepsilon_0 , \quad W_{pO} = 18(\frac{1}{2}\alpha^2 + \alpha)\, \sigma_0\varepsilon_0$$

Prob. 1.4 Same material as in Prob. 1.1, a material element undergoes the following uniaxial strain history: $\varepsilon = 0 \rightarrow \varepsilon_1 \rightarrow 0 \rightarrow -\varepsilon_1 \rightarrow 0$. The initial loading ends at $\varepsilon = \varepsilon_1$, $\sigma = (1.0 + \alpha)\, \sigma_0, (0 \leq \alpha \leq 1.0)$. Determine the stress history corresponding to the given strain history. Sketch the stress vs. strain curve. Assume isotropic hardening rule.

Solution: This problem is similar to Problem 1.1 except that a strain history instead of a stress history is given. The same solution procedure as in Problem 1.1 can be used.

The complete stress-strain history is plotted as solid curve in Fig. S1.4 for $\alpha = 0.4$. The material is elastic until the stress reaches the initial yield point, Point A, where $\sigma_A = \sigma_0$, $\varepsilon_A = \varepsilon_0 = \sigma_0 / E$. From the initial yield point to Point B where $\varepsilon_B = \varepsilon_1$, the material is in a plastic loading. From the result obtained in Problem 1.1, we find at $\sigma_B = (1 + \alpha)\, \sigma_0$, $\varepsilon_B = \varepsilon_1 = (1 + 10\alpha)\, \varepsilon_0$.

According to isotropic hardening rule, from $\varepsilon = \varepsilon_B$ to $\varepsilon = 0$, the material is in unloading and reversed elastic loading until it yields again in compression at Point D_1 where $\sigma_{D_1} = -\sigma_B = -(1 + \alpha)\, \sigma_0$. At this state, we have

$$\varepsilon_{D_1} = \varepsilon_B - 2\sigma_B / E = (1 + 10\alpha)\, \varepsilon_0 - 2(1 + \alpha)\, \sigma_0 / E = (8\alpha - 1)\, \varepsilon_0$$

From $\varepsilon = \varepsilon_{D_1}$, Point D_1, to $\varepsilon = 0$, Point E_1, the material is in a plastic loading, we obtain

$$\Delta\varepsilon = -(8\alpha - 1)\, \varepsilon_0 , \quad \Delta\sigma = E_t \Delta\varepsilon = -\frac{1}{10}(8\alpha - 1)\, \sigma_0$$

$$\sigma_{E_1} = \sigma_{D_1} + \Delta\sigma = -(0.9 + 1.8\alpha)\, \sigma_0$$

From $\varepsilon = 0$, Point E_1, to $\varepsilon = -(1 + 10\alpha)\, \varepsilon_0$, Point F_1, the material remains in a plastic loading state, and we obtain at Point F_1, $\varepsilon_{F_1} = -(1 + 10\alpha)\, \varepsilon_0$, $\sigma_{F_1} = -(1 + 2.8\alpha)\, \sigma_0$.

From $\varepsilon = \varepsilon_{F_1}$ to $\varepsilon = 0$, the material is first in a unloading state and then in a reversed elastic loading state, and finally in a plastic loading starting from Point G_1 where $\sigma_{G_1} = (1+2.8\alpha)\,\sigma_0$, according to the isotropic hardening rule. At $\sigma = \sigma_{G_1}$, we obtain

$$\varepsilon_{G_1} = \varepsilon_{F_1} + 2\,\sigma_{F_1}/E = -(1+10\alpha)\,\varepsilon + 2(1+2.8\alpha)\,\sigma_0/E = (1-4.4\alpha)\,\varepsilon_0$$

From Point G_1 to Point H_1 where $\varepsilon = 0$, the material is in a plastic loading state, and we determine the stress at Point H_1 as

$$\sigma_{H_1} = \sigma_{G_1} - E_t\,\varepsilon_{G_1} = (1+2.8\alpha)\,\sigma_0 + E_t(4.4\alpha - 1)\,\varepsilon_0 = (0.9 + 3.24\alpha)\,\sigma_0$$

The complete stress-strain history is given below.

$$[\varepsilon\,,\sigma] = [0\,,0]_O \to [\varepsilon_0\,,\sigma_0]_A \to [(1+10\alpha)\,\varepsilon_0\,,(1+\alpha)\,\sigma_0]_B \to$$
$$[(8\alpha-1)\,\varepsilon_0\,,-(1+\alpha)\,\sigma_0]_{D_1} \to [0\,,-(0.9+1.8\alpha)\,\sigma_0]_{E_1} \to$$
$$[-(1+10\alpha)\,\varepsilon_0\,,-(1+2.8\alpha)\,\sigma_0]_{F_1} \to$$
$$[(1-4.4\alpha)\,\varepsilon_0\,,(1+2.8\alpha)\,\sigma_0]_{G_1} \to [0\,,(0.9+3.24\alpha)\,\sigma_0]_{H_1}$$

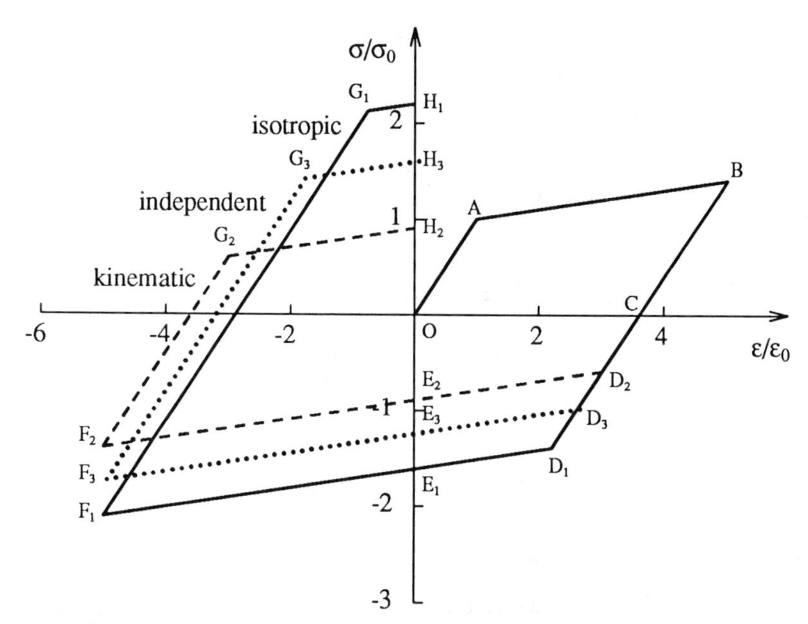

Figure S1.4. Stress vs. strain relations for Prob. 1.4 to Prob. 1.6

Prob. 1.5 The same stress-strain response problem as Prob. 1.4, except that the material is assumed to follow kinematic hardening rule.

Answer: The complete stress-strain history is

$$[\varepsilon , \sigma] = [0 , 0]_O \rightarrow [\varepsilon_0 , \sigma_0]_A \rightarrow [(1 + 10\alpha)\, \varepsilon_0 , (1 + \alpha)\, \sigma_0]_B \rightarrow$$
$$[(10\alpha - 1)\, \varepsilon_0 , -(1 - \alpha)\, \sigma_0]_{D_2} \rightarrow [0 , -0.9\, \sigma_0]_{E_2} \rightarrow$$
$$[-(1 + 10\alpha)\, \varepsilon_0 , -(1 + \alpha)\, \sigma_0]_{F_2} \rightarrow$$
$$[(1 - 10\alpha)\, \varepsilon_0 , (1 - \alpha)\, \sigma_0]_{G_2} \rightarrow [0 , 0.9\, \sigma_0]_{H_2}$$

and the stress vs. strain curve is plotted as dashed curve in Fig. S1.4.

Prob. 1.6 The same stress-strain response problem as Prob. 1.4, except that the material is assumed to follow independent hardening rule.

Answer: The complete stress-strain history is

$$[\varepsilon , \sigma] = [0 , 0]_O \rightarrow [\varepsilon_0 , \sigma_0]_A \rightarrow [(1 + 10\alpha)\, \varepsilon_0 , (1 + \alpha)\, \sigma_0]_B \rightarrow$$
$$[(9\alpha - 1)\, \varepsilon_0 , -\sigma_0]_{D_3} \rightarrow [0 , -(0.9 + 0.9\alpha)\, \sigma_0]_{E_3} \rightarrow$$
$$[-(1 + 10\alpha)\, \varepsilon_0 , -(1 + 1.9\alpha)\, \sigma_0]_{F_3} \rightarrow$$
$$[(1 - 7.1\alpha)\, \varepsilon , (1 + \alpha)\, \sigma_0]_{G_3} \rightarrow [0 , (0.9 + 1.71\alpha)\, \sigma_0]_{H_3}$$

and the stress vs. strain curve is plotted as dotted curve in Fig. S1.4.

1.6 Fixed-End Bar Problems

Prob. 1.7 A bar with fixed ends is subjected to a pair of equal and opposite axial forces. Both distances to the left end and to the right end are equal to a, and the distance between the two load points are equal to b, as shown in Fig. P1.7. The bar is made of an elastic-perfectly plastic material with yield stress σ_y. The axial force P is first increased from zero until the plastic flow occurs in the entire bar, and then P is unloaded to zero, followed by a reloading in the reverse direction. Assume b > 2a.

a. Determine the elastic and plastic limit loads P_e and P_p during the loading.
b. Determine the residual stress and plastic strain in the bar when the axial load P is unloaded to zero.
c. Determine the plastic limit load P_p during the reversed loading.
d. Sketch the P vs. u curve for the complete load-reversed load cycle for the case b = 3a, where u is the axial displacement of the bar at the left load point.

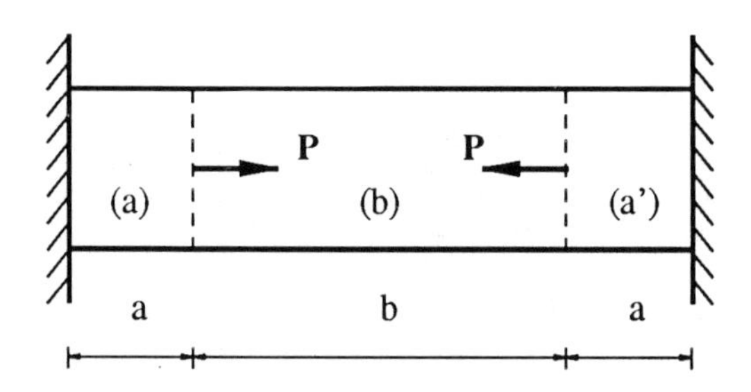

Figure P1.7.

Solution: Because of symmetry, we only need consider the stress and strain in the left Part a and in the Part b. The equilibrium and compatibility equations are

$$\sigma_a - \sigma_b = \frac{P}{A}$$

$$2a\,\varepsilon_a + b\,\varepsilon_b = 0$$

where σ_a and ε_a are stress and strain in Part a, and σ_b and ε_b are stress and strain in Part b respectively. A is area of the cross section. Using the elastic stress-strain relation, we obtain

$$\sigma_a = \frac{b}{L}\frac{P}{A}, \quad \sigma_b = -\frac{2a}{L}\frac{P}{A}$$

where $L = 2a + b$ is the length of the bar.

(a) *Elastic limit load* P_e *and plastic limit load* P_p

Because $b > 2a$, we note $|\sigma_a| > |\sigma_b|$, and plastic flow occurs first in Part a. Let $\sigma_a = \sigma_y$, we obtain the elastic limit load P_e

$$\frac{P_e}{A} = \frac{L}{b}\sigma_y > \sigma_y$$

For $P > P_e$, $\sigma_a = \sigma_y$, from the equilibrium equation, we have

$$\sigma_b = \sigma_y - \frac{P}{A}$$

Let $\sigma_b = -\sigma_y$, we obtain the plastic limit load P_p

$$\frac{P_p}{A} = 2\,\sigma_y$$

(b) *Unloading*

When P is decreased from P_p to zero, the material is in an elastically unloaded state, the incremental stresses are related to the incremental load ΔP by the elastic solutions

$$\Delta\sigma_a = \frac{b}{L}\frac{\Delta P}{A}, \quad \Delta\sigma_b = -\frac{2a}{L}\frac{\Delta P}{A}$$

Thus, we have

$$\sigma_a = \sigma_y + \frac{b}{L}\frac{\Delta P}{A}, \quad \sigma_b = -\sigma_y - \frac{2a}{L}\frac{\Delta P}{A}$$

Let $\Delta P = -P_p$, we obtain the residual stresses σ_a^* and σ_b^* in Part a and Part b

$$\sigma_a^* = \sigma_b^* = (1 - \frac{2b}{L})\,\sigma_y = -(1 - \frac{4a}{L})\,\sigma_y < 0$$

The residual stress is compressive, and $|\sigma_a| < \sigma_y$, no reversed yielding occurs during unloading.

The plastic strain in the bar for $P = 0$ is the same as for $P = P_p$, and we have, at $P = P_p$,

$$\varepsilon_b^p = 0, \quad \varepsilon_b = \varepsilon_b^e = -\frac{\sigma_y}{E} = -\varepsilon_y$$

$$\varepsilon_a = -\frac{b}{2a}\,\varepsilon_b = \frac{b}{2a}\,\varepsilon_y = \varepsilon_y + (\frac{b}{2a} - 1)\,\varepsilon_y$$

from which we obtain

$$\varepsilon_a^p = (\frac{b}{2a} - 1)\,\varepsilon_y$$

(c) *Reversed loading*

During the reversed loading, we have

$$\sigma_a = \sigma_a^* + \Delta\sigma_a = (1 - \frac{2b}{L})\,\sigma_y - \frac{b}{L}\frac{P'}{A}$$

$$\sigma_b = \sigma_b^* + \Delta\sigma_b = (1 - \frac{2b}{L})\,\sigma_y + \frac{2a}{L}\frac{P'}{A}$$

where $\Delta\sigma_a$ and $\Delta\sigma_b$ are the incremental stresses related to the reversed load P' which is increased from zero. Since we still have $|\sigma_a| > |\sigma_b|$, it follows that Part a of the bar yields first during the reversed loading. Let $\sigma_a = -\sigma_y$, we obtain in the reversed loading,

$$\frac{P'}{A} = \frac{P'_e}{A} = \frac{4a}{b}\,\sigma_y$$

Part a yields in compression. For $P' > P'_e$, using the equilibrium equation, we can again obtain the plastic limit load, P_p, in the reversed loading

$$\frac{P'_p}{A} = 2\,\sigma_y$$

(d) *Load and displacement relation*

For the case of b = 3a, the P vs. u relation can be obtained as

$$[\frac{P}{A\sigma_y}, \frac{u}{a\varepsilon_y}] = [0, 0] \rightarrow [\frac{5}{3}, 1] \rightarrow [2, \frac{3}{2}] \rightarrow [0, \frac{3}{10}] \rightarrow$$

$$[-\frac{4}{3}, -\frac{1}{2}] \rightarrow [-2, -\frac{3}{2}]$$

The relationship is shown as solid curve in Fig. S1.7.

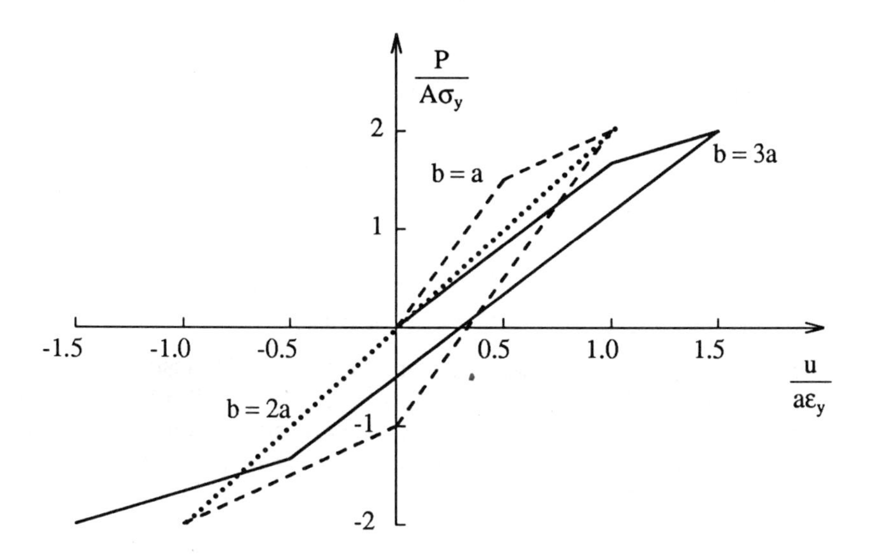

Figure S1.7. Load vs. displacement curves for Prob. 1.7 and Prob. 1.8

Prob. 1.8 Resolve the fixed-end bar problem in Prob. 1.7 for the cases b = a and b = 2a, respectively.

Hint: For the case b = a, Part b yields first. For the case b = 2a, Part a and Part b yield together.

Answer: For the case b = a (dashed curve in Fig. S1.7),

(a) $\frac{P_e}{A} = \frac{L}{2a}\,\sigma_y = \frac{3}{2}\,\sigma_y\,, \quad \frac{P_p}{A} = 2\,\sigma_y$

(b) $\sigma_a^* = \sigma_b^* = (1 - \frac{2b}{L})\,\sigma_y > 0\,, \quad \varepsilon_a^p = 0\,, \quad \varepsilon_b^p = -\,(1 + \frac{2a}{b})\,\varepsilon_y < 0$

(c) $\frac{P_e'}{A} = \frac{b}{a}\,\sigma_y = \sigma_y\,, \quad \frac{P_p'}{A} = 2\,\sigma_y$

(d) $[\frac{P}{A\sigma_y}\,,\,\frac{u}{a\varepsilon_y}] = [0\,,\,0] \to [\frac{3}{2}\,,\,\frac{1}{2}] \to [2\,,\,1] \to [-\,1\,,\,0] \to [-\,2\,,\,-\,1]$

For the case b = 2a (dotted curve in Fig. S1.7)

(a) $P_e = P_p = 2\,\sigma_y$

(b) $\sigma_a^* = \sigma_b^* = 0\,, \quad \varepsilon_a^p = \varepsilon_b^p = 0$

(c) $P_e' = P_p' = 2\,\sigma_y$

(d) $[\frac{P}{A\sigma_y}\,,\,\frac{u}{a\varepsilon_y}] = [0\,,\,0] \to [2\,,\,1] \to [0\,,\,0] \to [-2\,,\,-1]$

Prob. 1.9 The same fixed-end bar as in Prob. 1.7. Assume a linear work-hardening model

$$\sigma = E\,\varepsilon\,, \qquad \varepsilon \le \varepsilon_0$$
$$\sigma = \sigma_0 + E_t(\varepsilon - \varepsilon_0)\,, \quad \varepsilon > \varepsilon_0\,, \quad \sigma_0 = E\varepsilon_0$$

where $E_t = E/5$ is the plastic tangential modulus. Assume isotropic hardening rule. The bar is first loaded from $P = 0$ to $P = 5\,A\sigma_0/2$, and then is completely unloaded to $P = 0$. Determine the residual stresses, residual strains, and plastic strains in the bar after the complete unloading. Assume $b = 3a$.

Solution:

Basic Equations

The equilibrium and compatibility equations are the same as in Prob. 1.7,

$$\sigma_a - \sigma_b = \frac{P}{A}\,, \qquad 2a\,\varepsilon_a + b\,\varepsilon_b = 0$$

For this problem, we also need the incremental form of these equations. Use $\Delta\sigma$, $\Delta\varepsilon$, and ΔP to denote the incremental stress, strain, and load, the incremental equilibrium and compatibility equations are

$$\Delta\sigma_a - \Delta\sigma_b = \frac{\Delta P}{A}\,, \qquad 2a\,\Delta\varepsilon_a + b\,\Delta\varepsilon_b = 0$$

Elastic Solution

The elastic solution is the same as in Prob. 1.1,

$$\sigma_a = \frac{b}{L}\frac{P}{A}, \quad \sigma_b = -\frac{2a}{L}\frac{P}{A}$$

Since b = 3a, Part a yields first, and the elastic limit load P_e is

$$\frac{P_e}{A} = \frac{L}{b}\sigma_0$$

and at $P = P_e$, we have

$$\sigma_a = \sigma_0 , \quad \sigma_b = -\frac{2a}{b}\sigma_0 = -\frac{2}{3}\sigma_0$$

For $P > P_e$

For $P > P_e$, in Part a, the material is in a plastic loading state, and the incremental stress-strain relation is

$$\Delta\sigma_a = E_t\Delta\varepsilon_a = \frac{E}{5}\Delta\varepsilon_a$$

In Part b, the material is in an elastic loading state, the elastic stress-strain relation, $\Delta\sigma_b = E\,\Delta\varepsilon_b$ still holds. Combining the incremental stress-strain relations with the incremental equilibrium and compatibility equations, we obtain the incremental stresses expressed in terms of the incremental load ΔP for $P > P_e$

$$\Delta\sigma_a = \frac{b}{10a+b}\frac{\Delta P}{A}, \quad \Delta\sigma_b = -\frac{10a}{10a+b}\frac{\Delta P}{A}$$

Thus, for $P > P_e$, the stresses are

$$\sigma_a = \sigma_0 + \frac{b}{10a+b}\frac{P-P_e}{A}, \quad \sigma_b = -\frac{2a}{b}\sigma_0 - \frac{10a}{10a+b}\frac{P-P_e}{A}$$

The above solution holds until $\sigma_b = -\sigma_0$. Let $\sigma_b = -\sigma_0$, we obtain the stresses at $P = P_1 = \frac{21}{10}A\sigma_0$,

$$\sigma_b = -\sigma_0 , \quad \sigma_a = \frac{11}{10}\sigma_0$$

For $P > P_1$

For $P > P_1$, both parts of the bar are in a plastic loading state. Combining the elastic-plastic incremental relations

$$\Delta\sigma_a = E_t\Delta\varepsilon_a = \frac{E}{5}\Delta\varepsilon_a , \quad \Delta\sigma_b = E_t\Delta\varepsilon_b = \frac{E}{5}\Delta\varepsilon_b$$

with the incremental equilibrium and compatibility equations, we obtain the incremental stresses for $P > P_1$ as

$$\Delta\sigma_a = \frac{b}{L}\frac{\Delta P}{A} = \frac{3}{5}\frac{\Delta P}{A}, \quad \Delta\sigma_b = -\frac{2a}{L}\frac{\Delta P}{A} = -\frac{2}{5}\frac{\Delta P}{A}$$

Denote $P_2 = \frac{5}{2} A\,\sigma_0$, the final load, and let

$$\Delta P = P_2 - P_1 = \frac{5}{2} A\sigma_0 - \frac{21}{10} A\sigma_0 = \frac{4}{10} A\sigma_0$$

we obtain the incremental stresses

$$\Delta\sigma_a = \frac{6}{25}\sigma_0, \quad \Delta\sigma_b = -\frac{4}{25}\sigma_0$$

Thus, the stresses in the bar at $P = P_2 = \frac{5}{2} A\sigma_0$ are obtained as

$$\sigma_a = \frac{11}{10}\sigma_0 + \frac{6}{25}\sigma_0 = \frac{67}{50}\sigma_0, \quad \sigma_b = -\sigma_0 - \frac{4}{25}\sigma_0 = -\frac{58}{50}\sigma_0$$

Unloading and Residual Stresses

During the unloading, from $P = \frac{5}{2} A\sigma_0$ to $P = 0$, the material in both parts are in an elastic unloading state. The incremental stresses during the unloading can be easily obtained

$$\Delta\sigma_a = \frac{b}{L}\frac{\Delta P}{A} = -\frac{3}{2}\sigma_0, \quad \Delta\sigma_b = -\frac{2a}{L}\frac{\Delta P}{A} = \sigma_0$$

where $\Delta P = -\frac{5}{2} A\sigma_0$. We then obtain the residual stresses

$$\sigma_a^* = \frac{67}{50}\sigma_0 + \Delta\sigma_a = -\frac{4}{25}\sigma_0 < 0, \quad \sigma_b^* = -\frac{58}{50}\sigma_0 + \Delta\sigma_b = -\frac{4}{25}\sigma_0 < 0$$

Plastic Strains and Residual Strains

Because the plastic modulus E_p is a constant, the plastic strains can be obtained by first subtracting σ_0 from the stresses at $P = \frac{5}{2} A\sigma_0$, and then dividing them by E_p,

$$\varepsilon_a^p = (\frac{67}{50}\sigma_0 - \sigma_0)/E_p = \frac{34}{25}\varepsilon_0, \quad \varepsilon_b^p = (-\frac{58}{50}\sigma_0 + \sigma_0)/E_p = -\frac{16}{25}\varepsilon_0$$

The residual strains consist of two components, the elastic component corresponding to the residual stresses, and the plastic strain component. Thus, the residual strains are obtained as

$$\varepsilon_a^* = \frac{\sigma_a^*}{E} + \varepsilon_a^p = \frac{6}{5}\varepsilon_0, \quad \varepsilon_b^* = \frac{\sigma_b^*}{E} + \varepsilon_b^p = -\frac{4}{5}\varepsilon_0$$

Prob. 1.10 Resolve the same fixed-end bar problem as Prob. 1.9 for the case b = a.

Answer:

$$\sigma_a^* = \sigma_b^* = \frac{4}{15}\sigma_0\,, \quad \varepsilon_a^p = \frac{2}{5}\varepsilon_0\,, \quad \varepsilon_b^p = -\frac{8}{5}\varepsilon_0\,,$$

$$\varepsilon_a^* = \frac{2}{3}\varepsilon_0\,, \quad \varepsilon_b^* = -\frac{4}{3}\varepsilon_0$$

Prob. 1.11 Continue the fixed-end bar problem, Prob. 1.9. After the bar is unloaded to P = 0, reload the bar in the reversed direction. Determine the load P at which the bar yields again. Assume

a. Isotropic hardening rule
b. Kinematic hardening rule
c. Independent hardening rule

Solution: In Prob. 1.9, we have obtained the stresses at $P = \frac{5}{2} A\sigma_0$

$$\sigma_a = \frac{67}{50}\sigma_0, \qquad \sigma_b = -\frac{58}{50}\sigma_0$$

and the residual stresses at P = 0

$$\sigma_a^* = -\frac{4}{25}\sigma_0, \qquad \sigma_b^* = -\frac{4}{25}\sigma_0,$$

From the elastic solution of the bar obtained previously in Prob. 1.9, the stresses in the bar during the reversed loading are expressed as

$$\sigma_a = -\frac{4}{25} + \frac{3}{5}\frac{\Delta P}{A}$$

$$\sigma_b = -\frac{4}{25} - \frac{2}{5}\frac{\Delta P}{A}$$

where ΔP is the load increment in the reversed direction.

To determine the load at which the bar yields again, these two expressions should be used to check against the yield stresses in Part a and Part b, respectively.

(a) For isotropic hardening rule

According to isotropic hardening rule, Part a yields when

$$\sigma_a = -\frac{67}{50}\sigma_0$$

or

$$-\frac{4}{25}\sigma_0 + \frac{3}{5}\frac{\Delta P}{A} = -\frac{67}{50}\sigma_0$$

Thus, we obtain the load increment as

$$\frac{\Delta P}{A} = -\frac{59}{30}\sigma_0$$

Part b yields when

$$\sigma_b = \frac{58}{50}\sigma_0$$

or

$$-\frac{4}{25}\sigma_0 - \frac{2}{5}\frac{\Delta P}{A} = \frac{58}{50}\sigma_0$$

The corresponding load increment is determined as

$$\frac{\Delta P}{A} = -\frac{33}{10}\sigma_0$$

Therefore, Part a yields first in the reversed loading at the load $P = -\frac{59}{30}A\sigma_0$.

(b) For kinematic hardening rule

According to kinematic hardening rule, Part a yields when

$$\sigma_a = \frac{67}{50}\sigma_0 - 2\sigma_0$$

The corresponding load increment is obtained as $\Delta P = -\frac{5}{6}\sigma_0$. Part b yields when

$$\sigma_b = 2\sigma_0 - \frac{58}{50}\sigma_0$$

The corresponding load increment is obtained as $\Delta P = -\frac{5}{2}\sigma_0$. Therefore, Part a yields first in the reversed loading at the load $P = -\frac{5}{6}A\sigma_0$.

(c) For independent hardening rule

According to independent hardening rule, Part a yields when $\sigma_a = -\sigma_0$, and Part b yields when $\sigma_b = \sigma_0$. Thus, following the same procedure as the first two cases, we obtain that Part a yields first at the load $P = -\frac{7}{5}\sigma_0$.

Prob. 1.12 Continue the fixed-end bar problem in Prob. 1.10. After the bar is unloaded to P = 0, reload the bar in the reversed direction. Determine the load P at which the bar yields again. Assume

a. Isotropic hardening rule
b. Kinematic hardening rule
c. Independent hardening rule

Answer: In all cases, Part b yields first.

$$P = -\frac{17}{10} A \sigma_0 \qquad \text{for isotropic hardening rule}$$

$$P = -\frac{1}{2} A \sigma_0 \qquad \text{for kinematic hardening rule}$$

$$P = -\frac{11}{10} A \sigma_0 \qquad \text{for independent hardening rule}$$

Prob. 1.13 A bar with fixed-ends is subjected to an axial force P at a distance a from the left end and a distance b from the right-end with a < b, as shown in Fig. P1.13. The bar is made of an elastic-perfectly plastic material with yield stress σ_y. The axial force is first increased from zero until plastic flow occurs in the entire bar, and then is unloaded to zero, followed by a reloading in the reversed direction.

a. Determine the elastic and plastic limit loads P_e and P_p during the loading.
b. Determine the residual stress and plastic strain in the bar when the axial load P is unloaded to zero.
c. Determine the plastic limit load P_p during the reversed loading.
d. Sketch the P vs. u curve for the complete load-reversed load cycle for the case b = 3a, where u is the axial displacement of the bar at the load point.

Answer:

$$\text{(a)} \ \frac{P_e}{A} = \frac{L}{b} \sigma_y \,, \quad \frac{P_p}{A} = 2\,\sigma_y$$

$$\text{(b)} \ \sigma_a^* = \sigma_b^* = (1 - \frac{2b}{L})\,\sigma_y = -\,(1 - \frac{2a}{L})\sigma_y < 0$$

$$\text{(c)} \ \frac{P_e'}{A} = -\frac{2a}{b} \sigma_y \,, \quad \frac{P_p'}{A} = 2\,\sigma_y$$

$$\text{(d)} \ [\frac{P}{A\sigma_y} \,, \frac{u}{a\varepsilon_y}] = [0\,,0] \rightarrow [\frac{4}{3}\,,1] \rightarrow [2\,,3] \rightarrow [-\frac{2}{3}\,,1] \rightarrow [-2\,,-3]$$

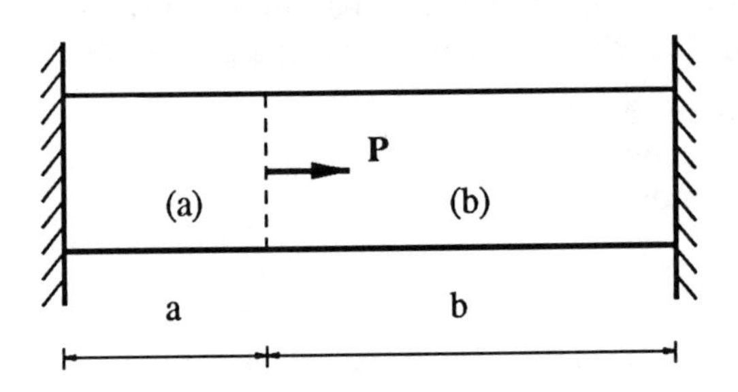

Figure P1.13.

The load-displacement curve is plotted in Fig. S1.13.

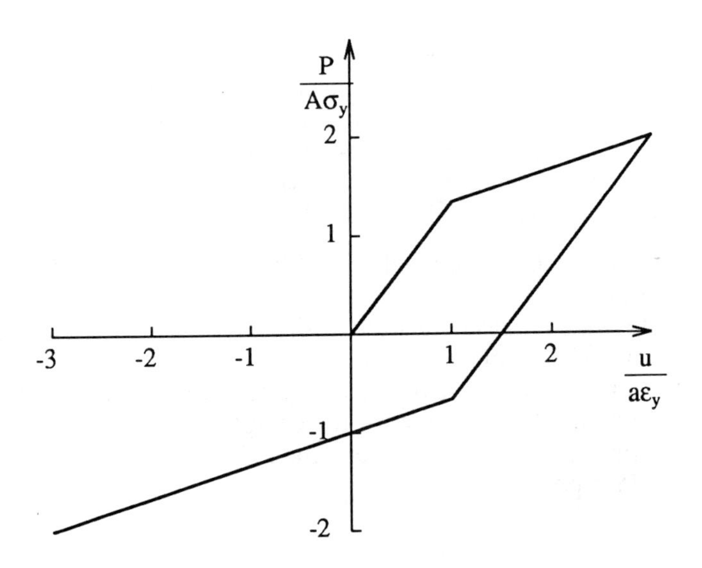

Figure S1.13. Load vs. displacement curves for Prob. 1.13

Prob. 1.14 The same fixed-end bar problem as in Prob. 1.13, assume the linear work-hardening model

$$\sigma = E\,\varepsilon\,, \qquad\qquad \varepsilon \le \varepsilon_0$$

$$\sigma = \sigma_0 + E_t(\varepsilon - \varepsilon_0), \quad \varepsilon > \varepsilon_0, \quad \sigma_0 = E\varepsilon_0$$

where $E_t = E/5$ is the plastic tangential modulus. Assume isotropic hardening rule. The bar is first loaded from $P = 0$ to $P = 3A\sigma_0$, and then is completely unloaded to $P = 0$. Determine the residual stress, residual strain, and plastic strain in the bar after the complete unloading. Assume $b = 3a$.

Answer:

$$\sigma_a^* = \sigma_b^* = -\frac{2}{5}\sigma_0, \quad \varepsilon_a^p = \frac{17}{5}\varepsilon_0, \quad \varepsilon_b^p = -\frac{3}{5}\varepsilon_0, \quad \varepsilon_a^* = 3\varepsilon_0, \quad \varepsilon_b^* = -\varepsilon_0$$

Prob. 1.15 Continue the fixed-end bar problem, Prob. 1.14. After the bar is unloaded to $P = 0$, reload the bar in the reversed direction. Determine the load P at which the bar yields again. Assume

a. Isotropic hardening rule
b. Kinematic hardening rule
c. Independent hardening rule

Answer: For the case of isotropic hardening rule, Part a yields first in the reversed loading at $P = -\frac{29}{15}A\sigma_0$.

For the case of kinematic hardening rule, Part a yields in compression during the unloading process at $P = \frac{1}{3}A\sigma_0$.

For the case of independent hardening rule, Part a yields in the reversed loading at $P = -\frac{4}{5}A\sigma_0$.

1.7 Stepped and Tapped Bar Problems

Prob. 1.16 A fixed-end bar with two areas of cross-section is shown in Fig. P1.16. Part a has an area A_1 and of length a, and Part b has an area A_2 and of length b. Assume $a > b$ and $A_2/A_1 > 1$. The bar is made of an elastic-perfectly plastic material with a yield stress σ_y, and is subjected to an axial force P at the intersection of the two parts. For the given loading path: $P = 0 \rightarrow P_p \rightarrow 0 \rightarrow P_p'$, where P_p and P_p' are the plastic limit loads in the initial loading and in the reversed loading, respectively

a. Determine P_e and P_p in the initial loading;

b. Determine the residual stresses and residual strains when P is unloaded to zero;
c. Determine the elastic limit load P_e' and the plastic limit load P_p' in the reversed loading;
d. For the case $A_2 / A_1 = 2$ and $a = 2b$, plot curves for P vs. u, σ_a vs. P, and σ_b vs. P.

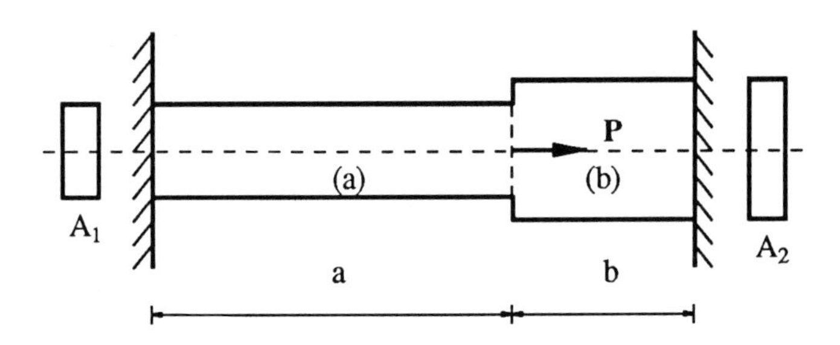

Figure P1.16.

Solution:

Basic Equations

The equilibrium and compatibility equations are

$$\sigma_a - \alpha \sigma_b = \frac{P}{A_1}, \qquad a\,\varepsilon_a + b\,\varepsilon_b = 0$$

where $\alpha = A_2 / A_1$.

Elastic Solution

Combining the above equations with the elastic stress-strain relation, the elastic solution for stresses is obtained as

$$\sigma_a = \frac{b}{b + \alpha a} \frac{P}{A_1}, \qquad \sigma_b = -\frac{a}{b + \alpha a} \frac{P}{A_1}$$

Initial Loading

Since $|\sigma_b| > |\sigma_a|$, the plastic flow starts in Part b at Point A (Fig. S1.16a). The elastic limit load is then obtained as

$$\frac{P_e}{A_1} = (\alpha + \frac{b}{a})\,\sigma_y$$

The plastic limit load at Point B is also easily obtained as

$$\frac{P_p}{A_1} = (1 + \alpha)\,\sigma_y$$

Unloading

Using the elastic solution, the residual stresses and strains after the load P is unloaded to zero are obtained as (Point C)

$$\sigma_a^* = \frac{\alpha\,(a-b)}{b+\alpha\,a}\,\sigma_y > 0, \qquad \sigma_b^* = \frac{(a-b)}{b+\alpha\,a}\,\sigma_y > 0$$

$$\varepsilon_a^p = 0, \qquad \varepsilon_b^p = -(\frac{a}{b} - 1)\,\varepsilon_y$$

$$\varepsilon_a^* = \frac{\alpha\,(a-b)}{b+\alpha\,a}\,\varepsilon_y > 0, \qquad \varepsilon_b^* = -\frac{\alpha\,a\,(a-b)}{b\,(b+\alpha\,a)}\,\varepsilon_y$$

Reversed Loading

Using the elastic solution and the expressions of residual stresses, stresses during the reversed loading are expressed as

$$\sigma_a = \frac{\alpha\,(a-b)}{b+\alpha\,a}\,\sigma_y + \frac{b}{b+\alpha\,a}\,\frac{\Delta P}{A_1}$$

$$\sigma_b = \frac{(a-b)}{b+\alpha\,a}\,\sigma_y - \frac{a}{b+\alpha\,a}\,\frac{\Delta P}{A_1}$$

where ΔP is the incremental load. From the above expressions, we see that Part b yields first in the reversed loading. The elastic limit load and the plastic limit load in reversed loading are obtained as

$$\frac{P_e'}{A_1} = (1 - \alpha - 2\,\frac{b}{a})\,\sigma_y \qquad \text{(Point D)}$$

$$\frac{P_p'}{A_1} = -(1 + \alpha)\,\sigma_y \qquad \text{(Point E)}$$

For $A_2 / A_1 = 2$ and $a = 2b$, the P vs. u curve is plotted in Fig. S1.16a, and the σ_a, σ_b vs. P curves are plotted in Fig. S1.16b.

Prob. 1.17 The same stepped bar as in Prob. 1.16. Assume a linear work-hardening model

$$\sigma = E\,\varepsilon\,, \qquad \sigma \le \sigma_0$$

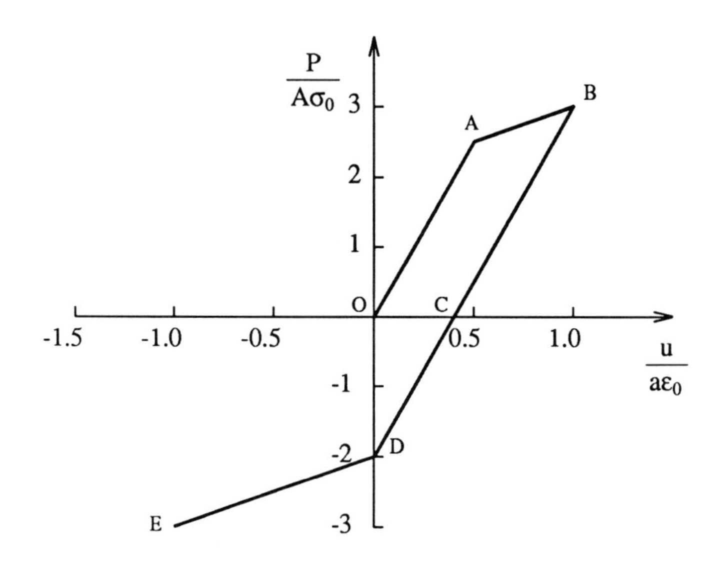

Figure S1.16a. Load vs. displacement curve for Prob. 1.16

$$\sigma = \sigma_0 + m\,\varepsilon^p\,, \qquad \sigma > \sigma_0$$

where $m = E/4$. Determine the stress history in both Part a and Part b for the loading path: $P = 0 \rightarrow 4\,A_1\sigma_0 \rightarrow 0 \rightarrow -4\,A_1\sigma_0 \rightarrow 0$. Assume $\alpha = A_2 / A_1 = 2$, $a/b = 2$. Assume isotropic hardening rule.

Solution: The basic equations are

$$\sigma_a - 2\,\sigma_b = \frac{P}{A_1}\,, \qquad \text{or} \qquad \Delta\sigma_a - 2\,\Delta\sigma_b = \frac{\Delta P}{A_1}$$

$$\varepsilon_a + \frac{1}{2}\,\varepsilon_b = 0\,, \qquad \text{or} \qquad \Delta\varepsilon_a + \frac{1}{2}\,\Delta\varepsilon_b = 0$$

Elastic Solution

The elastic solution is obtained as

$$\sigma_a = \frac{1}{5}\,\frac{P}{A_1}\,, \qquad \sigma_b = -\frac{2}{5}\,\frac{P}{A_1}$$

It is clear that Part b yields first, and the elastic limit load is

$$\frac{P_1}{A_1} = \frac{P_e}{A_1} = \frac{5}{2}\,\sigma_0$$

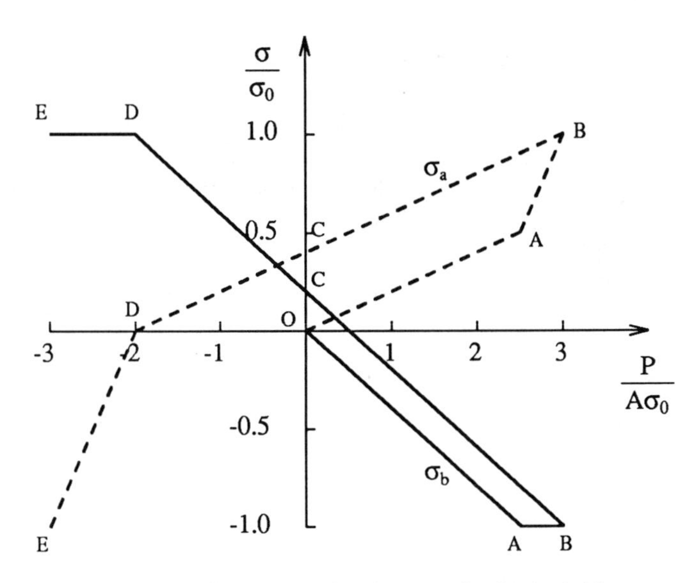

Figure S1.16b. Stresses vs. load curves for Prob. 1.16

and at this load, $\sigma_a = \dfrac{1}{2}\sigma_0$, $\sigma_b = -\sigma_0$.

For $P > P_1$

For $P > P_e$, the elastic-plastic stress-strain relation is used for stress in Part b

$$\Delta\sigma_b = E_t \Delta\varepsilon = \frac{E}{5}\Delta\varepsilon_b$$

Thus, combining the stress-strain relations with the basic equations leads to

$$\Delta\sigma_a - 2\,\Delta\sigma_b = \frac{\Delta P}{A_1}, \qquad \Delta\sigma_a + \frac{5}{2}\Delta\sigma_b = 0$$

from which, we obtain

$$\Delta\sigma_a = \frac{5}{9}\frac{\Delta P}{A_1}, \qquad \Delta\sigma_b = -\frac{2}{9}\frac{\Delta P}{A_1}$$

Let $\sigma_a = \sigma_0$, we obtain at $P = P_2 = \dfrac{34}{10} A_1\sigma_0$, $\sigma_a = \sigma_0$ and $\sigma_b = -\dfrac{6}{5}\sigma_0$.

For $P > P_2$

For $P > P_2$, the elastic-plastic stress-strain relation applies to both parts of the bar. Combining this with the basic equations, we obtain

$$\Delta\sigma_a - 2\,\Delta\sigma_b = \frac{\Delta P}{A_1}\,, \qquad \Delta\sigma_a + \frac{1}{2}\,\Delta\sigma_b = 0$$

The solution of these equations is

$$\Delta\sigma_a = \frac{1}{5}\frac{\Delta P}{A_1}\,, \qquad \Delta\sigma_b = -\frac{2}{5}\frac{\Delta P}{A_1}$$

Let $\frac{\Delta P}{A_1} = \frac{6}{10}\,\sigma_0$, the stress increments can be determined. Thus, we obtain at $P = P_3 = 4\,A_1\sigma_0$,

$$\sigma_a = \frac{28}{25}\,\sigma_0\,, \qquad \sigma_b = -\frac{36}{25}\,\sigma_0$$

Unloading to $P = 0$

Utilizing the elastic solution, the residual stresses at $P = P_4 = 0$ are obtained as

$$\sigma_a^* = \frac{8}{25}\,\sigma_0\,, \qquad \sigma_b^* = \frac{4}{25}\,\sigma_0$$

Reversed Loading

For $P < 0$, the material behaves elastically. The elastic stresses are

$$\sigma_a = \frac{8}{25}\,\sigma_0 + \frac{1}{5}\frac{\Delta P}{A_1}\,, \qquad \sigma_b = \frac{4}{25}\,\sigma_0 - \frac{2}{5}\frac{\Delta P}{A_1}\,,$$

According to isotropic hardening rule, Part a yields at $\sigma_a = -\frac{28}{25}\,\sigma_0$, and Part b yields at $\sigma_b = \frac{36}{25}\,\sigma_0$ in the reversed loading. Since Part b yields first in the reversed loading, we obtain at $P = P_5 = -\frac{16}{5}\,A_1\sigma_0$,

$$\sigma_a = \frac{-8}{25}\,\sigma_0\,, \qquad \sigma_b = \frac{36}{25}\,\sigma_0$$

For $P < P_5$

For $P < P_5$ until $P = P_6 = -4\,A_1\sigma_0$, Part a remains in elastic state. The elastic-plastic stress-strain relation is used for stress in Part b. Thus, we obtain at $P = P_6 = -4\,A_1\sigma_0$

$$\sigma_a = -\frac{172}{225}\,\sigma_0\,, \qquad \sigma_b = \frac{364}{225}\,\sigma_0$$

Unloading to $P = 0$

During the unloading from $P = P_6 = -4\,A_1\sigma_0$ to $P = P_7 = 0$, the material behaves elastically. Utilizing the elastic solution, the residual stresses at $P = 0$

are obtained as

$$\sigma_a = \frac{8}{225}\sigma_0\,, \qquad \sigma_b = \frac{4}{225}\sigma_0$$

The stress histories for Part a and Part b are summarized as:

$$\frac{P}{A_1\sigma_0} = [0,\ \frac{5}{2},\ \frac{34}{10},\ 4,\ 0,\ -\frac{16}{5},\ -4,0]$$

$$\frac{\sigma_a}{\sigma_0} = [0,\ \frac{1}{2},\ 1,\ \frac{28}{25},\ \frac{8}{25},\ -\frac{8}{25},\ -\frac{172}{225},\ \frac{8}{225}]$$

$$\frac{\sigma_b}{\sigma_0} = [0,\ -1,\ -\frac{6}{5},\ -\frac{36}{25},\ \frac{4}{25},\ \frac{36}{25},\ \frac{364}{225},\ \frac{4}{225}]$$

Prob. 1.18 The same stepped bar problem as Prob. 1.17. Assume kinematic hardening rule.

Answer: The stress histories for Part a and Part b are found as:

$$\frac{P}{A_1\sigma_0} = [0,\ \frac{5}{2},\ \frac{34}{10},\ 4,\ 0, -1,\ -\frac{14}{5},\ -4,\ 0]$$

$$\frac{\sigma_a}{\sigma_0} = [0,\ \frac{1}{2},\ 1,\ \frac{28}{25},\ \frac{8}{25},\ \frac{3}{25},\ -\frac{22}{25},\ -\frac{28}{25},\ -\frac{8}{25}]$$

$$\frac{\sigma_b}{\sigma_0} = [0,\ -1,\ -\frac{6}{5},\ -\frac{36}{25},\ \frac{4}{25},\ \frac{14}{25},\ \frac{24}{25},\ \frac{36}{25},\ -\frac{4}{25}]$$

Prob. 1.19 The same stepped bar problem as Prob. 1.17. Assume independent hardening rule.

Answer: The stress histories for Part a and Part b are found as:

$$\frac{P}{A_1\sigma_0} = [0,\ \frac{5}{2},\ \frac{34}{10},\ 4,\ 0, -\frac{21}{10},\ -\frac{186}{50},\ -4,\ 0]$$

$$\frac{\sigma_a}{\sigma_0} = [0,\ \frac{1}{2},\ 1,\ \frac{28}{25},\ \frac{8}{25},\ -\frac{1}{10},\ -1,\ -\frac{264}{250},\ -\frac{64}{250}]$$

$$\frac{\sigma_b}{\sigma_0} = [0,\ -1,\ -\frac{6}{5},\ -\frac{36}{25},\ \frac{4}{25},\ 1,\ \frac{34}{25},\ \frac{368}{250},\ -\frac{32}{250}]$$

Prob. 1.20 An one-end-fixed bar with a linearly-varying area is subjected to an axial force at its free end, Fig. P1.20. The areas at the two ends are A_1 and A_2 respectively, and $A_2 / A_1 = 2$. The bar is made of an elastic-linear work hardening material, and the stress-strain relationship in simple tension is expressed as

$$\sigma = E\,\varepsilon\,, \qquad \sigma \le \sigma_0$$

$$\sigma = \sigma_0 + m\,\varepsilon^p\,, \qquad \sigma > \sigma_0$$

where $m = E/5$. The load P is first increased from zero to $2A_1\sigma_0$, and then is released to zero.

a. Determine the elastic limit load P_e.
b. Determine the stress, strain, and plastic strain distribution in the bar for $P > P_e$, and sketch σ vs. x and ε vs. x curves.
c. Determine the residual stress and strain.
d. Sketch the P vs. u curve where u is the free end displacement.

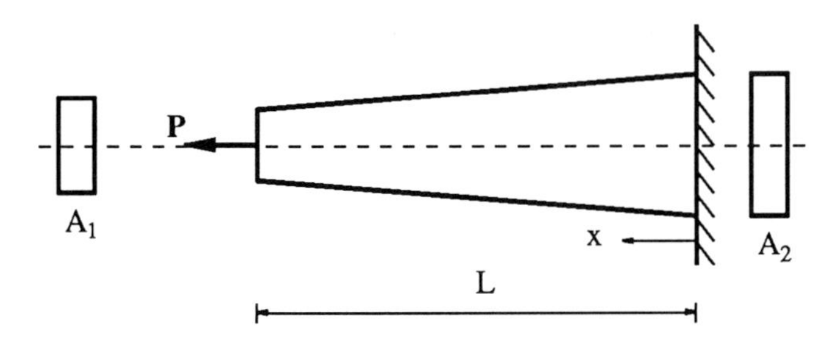

Figure P1.20.

Solution: Let $\bar{x} = x/L$. The area can be expressed as $A = (2 - \bar{x})\,A_1$.

Elastic Solution

The stress in the bar is statically determinate and therefore can be determined independent of the stress-strain relation. It has the following distribution

$$\sigma(\bar{x}) = \frac{P}{A} = \frac{P}{A_1(2-\bar{x})}$$

The maximum stress occurs at $\bar{x} - 1$. The elastic limit load P_e is obtained as

$$\frac{P_e}{A_1} = \sigma_0$$

Strain Distribution

The elastic strain distribution can be obtained from the stress expression as

$$\varepsilon^e(\bar{x}) = \frac{\sigma}{E} = \frac{1}{2-\bar{x}}\frac{P}{EA_1}$$

For $P > P_e$, there exists an elastic and plastic boundary point $\bar{x}_0$ such that for $0 \le \bar{x} \le \bar{x}_0$, the material is in the elastic state, and for $\bar{x}_0 < x < 1$, the material is in the elastic-plastic state. Let $\sigma(\bar{x}) = \sigma_0$, we obtain the boundary point as

$$\bar{x}_0 = 2 - \frac{P}{P_e}$$

In the elastic-plastic region ($\bar{x} > \bar{x}_0$), the plastic strain distribution is

$$\varepsilon^p(\bar{x}) = (\sigma - \sigma_0)/m = \frac{P_e}{mA_1}[\frac{1}{2-\bar{x}}\frac{P}{P_e} - 1]$$

Thus, the total strain distribution has the form

$$\varepsilon(\bar{x}) = \frac{1}{2-\bar{x}}\frac{P}{EA_1} \qquad \text{for } 0 \le \bar{x} \le \bar{x}_0$$

$$\varepsilon(\bar{x}) = \frac{1}{2-\bar{x}}\frac{P}{EA_1} + \frac{P_e}{mA_1}[\frac{1}{2-\bar{x}}\frac{P}{P_e} - 1] \qquad \text{for } \bar{x}_0 < \bar{x}$$

The stress and strain distributions for $\bar{x}_0 = 0.5$ are plotted in Fig. S1.20a.

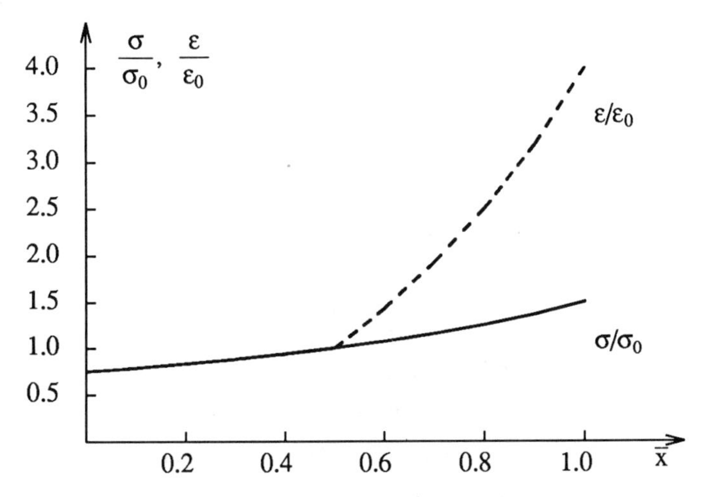

Figure S1.20a. Stress and strain distributions for $\bar{x}_0 = 0.5$

Free End Displacement

The free end displacement is found by integrating the strain over the length of the bar as

$$u = \int_0^L \varepsilon(x)\,dx = L\int_0^1 \varepsilon^e(\bar{x})\,d\bar{x} + L\int_0^1 \varepsilon^p(\bar{x})\,d\bar{x} = u^e + u^p$$

from which we obtain

$$u^e = L\int_0^1 \varepsilon^e(\bar{x})\,d\bar{x} = \frac{PL}{EA_1}(ln\,2)$$

and

$$u^p = L\int_0^1 \varepsilon^p(\bar{x})\,d\bar{x} = L\int_{\bar{x}_0}^1 \varepsilon^p(\bar{x})\,d\bar{x} = \frac{P_eL}{mA_1}\left[\frac{P}{P_e}\,ln\,\frac{P}{P_e} - \frac{P}{P_e} + 1\right]$$

Thus, the free end displacement has the value

$$u = \frac{PL}{EA_1}(ln\,2), \qquad \text{for } P \le P_e$$

$$u = \frac{PL}{EA_1}(ln\,2) + \frac{P_eL}{mA_1}\left[\frac{P}{P_e}\,ln\,\frac{P}{P_e} - \frac{P}{P_e} + 1\right], \qquad \text{for } P > P_e$$

The P vs. u curve is plotted as solid curve in Fig. S1.20b.

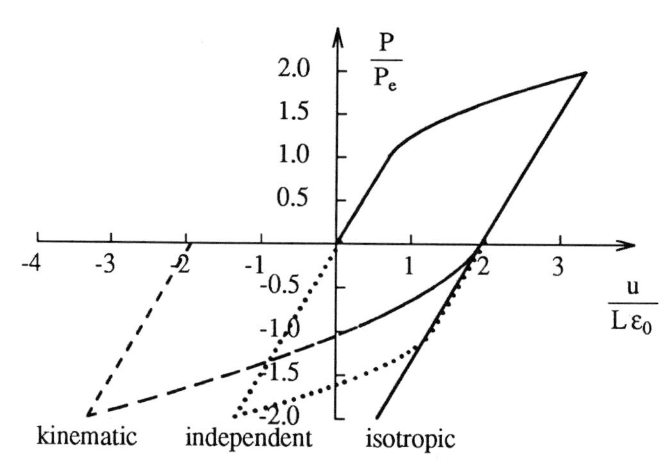

Figure S1.20b. P vs. u curves for Probs. 1.20, 1.21, 1.22, and 1.23

Unloading

At $P = 2\,A_1\sigma_0$, we have

$$\bar{x}_0 = 0, \qquad \sigma = \frac{2}{2-\bar{x}}\sigma_0$$

$$\varepsilon^p = \frac{P_e}{mA_1}\frac{\bar{x}}{2-\bar{x}} \qquad \varepsilon = \frac{2}{2-\bar{x}}\varepsilon_0 + \frac{P_e}{mA_1}\frac{\bar{x}}{2-\bar{x}}$$

When the bar is unloaded to $P = 0$, we have

$$\sigma = 0, \qquad \varepsilon = \varepsilon^p = \frac{P_e}{mA_1}\frac{\bar{x}}{2-\bar{x}}$$

and

$$u = u^p = \frac{P_e L}{mA_1}(\ln 4 - 1)$$

Prob. 1.21 Continue the tapped bar problem, Prob. 1.20. After the load P is released to zero, reload the bar in the reversed direction to the load $-2A_1\sigma_0$, and then release the load again. Sketch the P vs. u curve for the reversed loading and unloading process. Assume isotropic hardening rule.

Hint: Since the plastic deformation is not uniformly distributed along the length of the bar, the yield stress in reversed direction is a function of $\bar{x}$. The distribution of the subsequent yielding stress in reversed loading is

$$\sigma_y = -\frac{2}{2-\bar{x}}\sigma_0$$

Answer: The entire bar remains in an elastic state during the reversed loading. The displacement during the reversed loading can be expressed as

$$u = \frac{P_e L}{mA_1}(\ln 4 - 1) - \frac{PL}{EA_1}(\ln 2)$$

where P is the absolute value of the reversed load. The P vs. u curve is plotted as solid curve in Fig. S1.20b.

Prob. 1.22 The same tapped bar problem as Prob. 1.21. Assume kinematic hardening rule.

Hint: The distribution of the subsequent yielding stress in reversed loading is

$$\sigma_y = \frac{2}{2-\bar{x}}\sigma_0 - 2\sigma_0 = -\frac{1-\bar{x}}{2-\bar{x}}2\sigma_0$$

Answer: The displacement during the reversed loading can be expressed as

$$u = \frac{P_e L}{m A_1}[ln 4 - 1] - \frac{PL}{EA_1} ln 2 - \frac{P_e L}{m A_1}[(2 + \frac{P}{P_e}) ln (1 + \frac{1}{2}\frac{P}{P_e}) - \frac{P}{P_e}]$$

where P is the absolute value of the reversed load. The P vs. u curve is plotted as dashed curve in Fig. S1.20b.

Prob. 1.23 The same tapped bar problem as Prob. 1.21. Assume independent hardening rule.

Answer: For $P > P_e'$ in reversed loading, the displacement can be expressed as

$$u = \frac{P_e L}{m A_1}[ln 4 - 1] - \frac{PL}{EA_1}(ln 2) - \frac{P_e L}{m A_1}[\frac{P}{P_e} ln \frac{P}{P_e} - \frac{P}{P_e} + 1]$$

where P is the absolute value of the reversed load. The P vs. u curve is plotted as dotted curve in Fig. S1.20b.

1.8 Three-Bar Structure Problems

Prob. 1.24 A structure consisting of three parallel vertical bars and a horizontal rigid bar is subjected a vertical load as shown in Fig. P1.24. Three vertical bars are made of elastic-perfectly plastic materials. The left and right bars (marked as Bar 1 in Fig. P1.24) have same area A_1 and same yield stress, σ_{10}. The middle bar (Bar 2 in Fig. P1.24) has an area A_2 and yield stress σ_{20}. Assume $\sigma_{10} \geq \sigma_{20}$. The load P is first increased to the plastic limit load P_p, and then is decreased to zero.

a. Determine the elastic limit load P_e and the plastic limit load P_p;
b. Determine the residual stresses and strains in the bars after the load is decreased to zero;
c. Determine the condition such that bar 2 will not yield again in the unloading.

Answer:

a. $\frac{P_e}{A_2} = (1 + 2\alpha)\sigma_{20}, \quad \frac{P_p}{A_2} = 2\alpha\sigma_{10} + \sigma_{20}, \quad \text{where } \alpha = A_1 / A_2.$

b. $\sigma_1^* = \frac{1}{1 + 2\alpha}(\sigma_{10} - \sigma_{20}), \quad \sigma_2^* = -\frac{2\alpha}{1 + 2\alpha}(\sigma_{10} - \sigma_{20})$

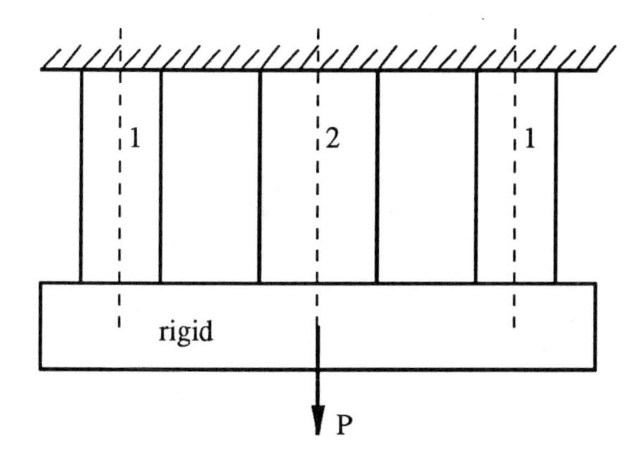

Figure P1.24.

c. $\sigma_{10} < \dfrac{1+4\alpha}{2\alpha}\sigma_{20}$.

Prob. 1.25 The same three-bar structure as in Prob. 1.24 except that all three bars have the same area A. Assume elastic-linear work hardening materials. The stress-strain relation in simple tension is expressed as

$$\sigma = E\,\varepsilon\,, \qquad \sigma \le \sigma_0$$

$$\sigma = \sigma_0 + m\,\varepsilon^p\,, \qquad \sigma > \sigma_0$$

The plastic modulus m is the same for all three bars, and $m = E/6$. However, the initial yield stress for the left and right bars is σ_{10} and for the middle bar is σ_{20}, and $\sigma_{10} = 2\sigma_{20}$. For the loading history: $P = 0 \rightarrow 6A\sigma_{20} \rightarrow 0 \rightarrow -6A\sigma_{20} \rightarrow 0$, sketch the σ_1 vs. v, σ_2 vs. v, and P vs. v curves., where v is the vertical displacement at the load point. Assume isotropic hardening rule.

Answer: The relationships between displacement, stress, and load are given below and plotted in Figs. S1.25a and S1.25b (solid curve) respectively.

$$\frac{v}{L\varepsilon_{20}} = [0,\ 1,\ 2,\ 4,\ 2,\ \frac{8}{7},\ -\frac{16}{35},\ \frac{54}{35}]$$

$$\frac{\sigma_1}{\sigma_{20}} = [0,\ 1,\ 2,\ \frac{16}{7},\ \frac{2}{7},\ -\frac{4}{7},\ -\frac{76}{35},\ -\frac{6}{35}]$$

$$\frac{\sigma_2}{\sigma_{20}} = [0,\ 1,\ \frac{8}{7},\ \frac{10}{7},\ -\frac{4}{7},\ -\frac{10}{7},\ -\frac{58}{35},\ -\frac{12}{35}]$$

$$\frac{P}{A\sigma_{20}} = [0,\ 3,\ \frac{36}{7},\ 6,\ 0,\ -\frac{18}{7},\ -6,\ 0]$$

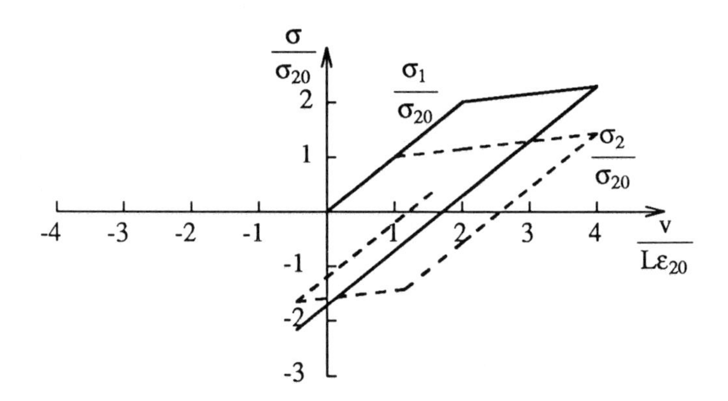

Figure S1.25a. Stress vs. displacement curves for Prob. 1.25

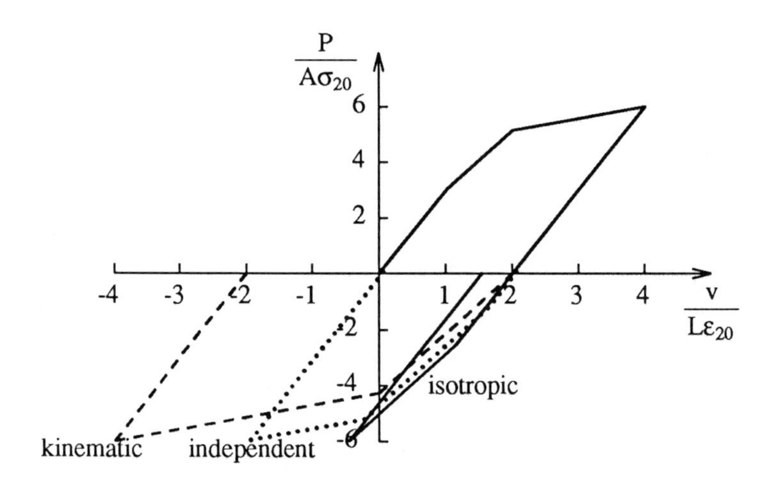

Figure S1.25b. Load vs. displacement curves for Probs. 1.25, 1.26 and 1.27

Prob. 1.26 The same three-bar problem as Prob. 1.25. Assume kinematic hardening rule.

Answer: The relationships between displacement, stress, and load are given below and plotted in Figs. S1.25b (dashed curve) and S1.26 respectively.

$$\frac{v}{L\varepsilon_{20}} = [0,\ 1,\ 2,\ 4,\ 2,\ 0,\ -4\ ,-2]$$

$$\frac{\sigma_1}{\sigma_{20}} = [0,\ 1,\ 2,\ \frac{16}{7},\ \frac{2}{7},\ -\frac{12}{7},\ -\frac{16}{7},\ -\frac{2}{7}]$$

$$\frac{\sigma_2}{\sigma_{20}} = [0,\ 1,\ \frac{8}{7},\ \frac{10}{7},\ -\frac{4}{7},\ -\frac{6}{7},\ -\frac{10}{7},\ \frac{4}{7}]$$

$$\frac{P}{A\sigma_{20}} = [0,\ 3,\ \frac{36}{7},\ 6,\ 0,\ -\frac{30}{7},\ -6,\ 0]$$

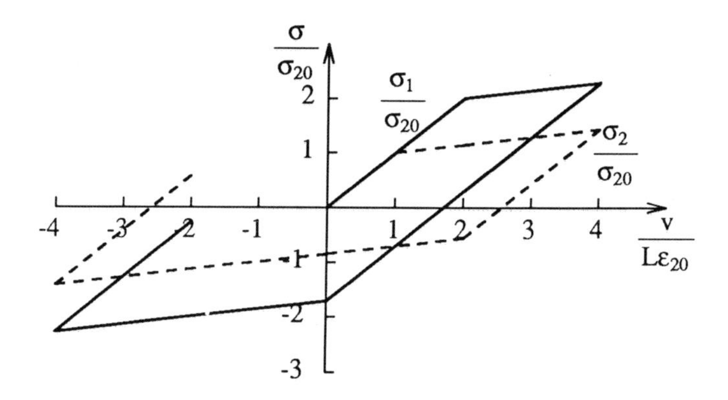

Figure S1.26. Stress vs. displacement curves for Prob. 1.26

Prob. 1.27 The same three-bar problem as Prob. 1.25. Assume independent hardening rule.

Answer: The relationships between displacement, stress, and load are given below and plotted in Fig. S1.25b (dotted curve) and Fig. S1.27 respectively.

$$\frac{v}{L\varepsilon_{20}} = [0,\ 1,\ 2,\ 4,\ 2,\ \frac{11}{7},\ -\frac{2}{7},\ -2,\ 0]$$

$$\frac{\sigma_1}{\sigma_{20}} = [0,\ 1,\ 2,\ \frac{16}{7},\ \frac{2}{7},\ -\frac{1}{7},\ -2,\ -\frac{110}{49},\ -\frac{12}{49}]$$

$$\frac{\sigma_2}{\sigma_{20}} = [0,\ 1,\ \frac{8}{7},\ \frac{10}{7},\ -\frac{4}{7},\ -1,\ -\frac{62}{49},\ -\frac{74}{49},\ \frac{24}{49}]$$

$$\frac{P}{A\sigma_{20}} = [0,\ 3,\ \frac{36}{7},\ 6,\ 0,\ -\frac{9}{7},\ -\frac{258}{49},\ -6,\ 0]$$

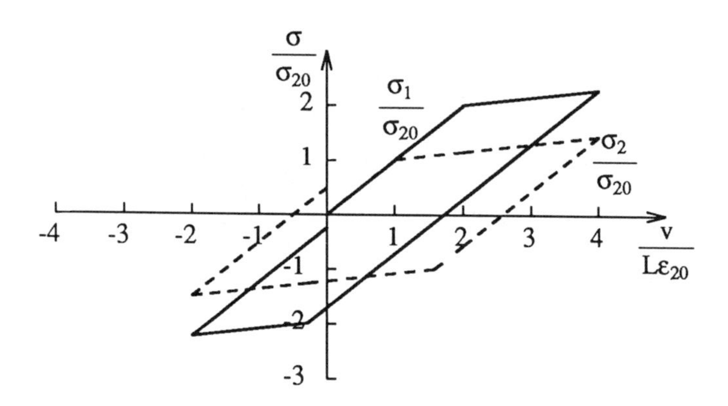

Figure S1.27. Stress vs. displacement curves for Prob. 1.27

1.9 Three-Bar Truss Problems

Prob. 1.28 A three-bar truss subjected to a vertical load P is shown in Fig. P1.28. The bars are made of an elastic-perfectly plastic material with elastic modulus E and yield stress σ_0. All three bars have the same area A. The load P is first increased to the plastic limit P_p and then is unloaded to zero. Afterward, P is increased in the reversed direction until all three bars yield again in compression and then is unloaded again to zero.

a. Determine the elastic limit load P_e and the plastic limit load P_p in the initial loading.
b. Determine the elastic limit load P_e' and the plastic limit load P_p' in the reversed loading.
c. Determine the residual stresses and strains of the bars and the residual horizontal and vertical displacements at the load point at the end of the loading history.

Solution: The basic equilibrium and compatibility equations are

$$\sigma_1 = \sigma_2, \qquad \sigma_1 + \sigma_3 = \frac{P}{A}$$

$$\varepsilon_1 = \varepsilon_2, \qquad 4\,\varepsilon_1 = \varepsilon_3$$

Using the elastic stress-strain relations, we obtain the elastic solution as

$$\sigma_1 = \sigma_2 = \frac{1}{5}\frac{P}{A}, \qquad \sigma_3 = \frac{4}{5}\frac{P}{A}$$

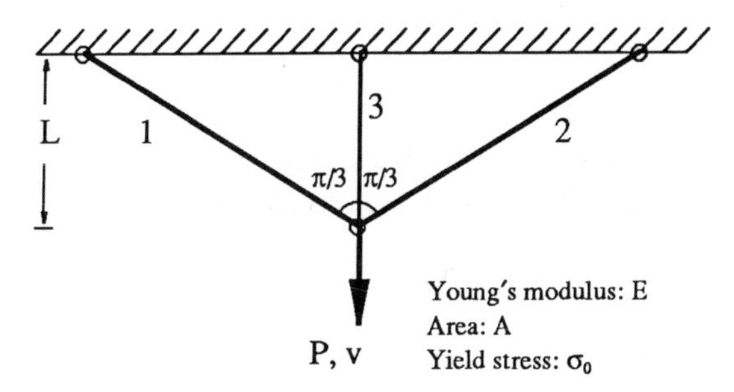

Figure P1.28.

Initial Loading

The elastic limit load is obtained by equating σ_3 to σ_0,

$$\frac{P_e}{A} = \frac{5}{4}\sigma_0$$

and at $P = P_e$, we have

$$\sigma_1 = \sigma_2 = \frac{1}{4}\sigma_0, \quad \sigma_3 = \sigma_0, \quad \varepsilon_1 = \varepsilon_2 = \frac{1}{4}\varepsilon_0, \quad \varepsilon_3 = \varepsilon_0$$

$$u = 0, \quad v = L\varepsilon_0$$

where u and v are the horizontal and vertical displacements respectively at the load point.

For $P > P_e$, $\Delta\sigma_3 = 0$. Thus, from the basic equations we obtain the plastic limit load as

$$\frac{P_p}{A} = 2\sigma_0$$

and at $P = P_p$, we have

$$\sigma_1 = \sigma_2 = \sigma_3 = \sigma_0, \quad \varepsilon_1 = \varepsilon_2 = \varepsilon_0, \quad \varepsilon_3 = 4\varepsilon_0$$

$$u = 0, \quad v = 4L\varepsilon_0$$

Unloading

When the load P is unloaded to zero, the incremental stresses are

$$\Delta\sigma_1 = \Delta\sigma_2 = -\frac{2}{5}\sigma_0$$

Thus, at $P = 0$, we obtain

$$\sigma_1 = \sigma_2 = \frac{3}{5}\sigma_0, \quad \sigma_3 = -\frac{3}{5}\sigma_0, \quad \varepsilon_1 = \varepsilon_2 = \frac{3}{5}\varepsilon_0, \quad \varepsilon_3 = \frac{12}{5}\varepsilon_0$$

$$u = 0, \qquad v = \frac{12}{5} L \varepsilon$$

Reversed Loading

In the reversed loading, Bar 3 yields first again. We can easily obtain the elastic limit and plastic limit load in the reversed loading.

$$\frac{P_e'}{A} = -\frac{1}{2}\sigma_0, \qquad \frac{P_p'}{A} = -2\sigma_0$$

and at $P = P_p'$, we have

$$\sigma_1 = \sigma_2 = \sigma_3 = -\sigma_0, \qquad \varepsilon_1 = \varepsilon_2 = -\varepsilon_0, \qquad \varepsilon_3 = -4\,\varepsilon_0$$

$$u = 0, \qquad v = -4L\varepsilon_0$$

Unloading

Finally, when the load P is unloaded to zero again, we obtain the residual stresses and strains as

$$\sigma_1^* = \sigma_2^* = -\frac{3}{5}\sigma_0, \quad \sigma_3^* = \frac{3}{5}\sigma_0, \qquad \varepsilon_1^* = \varepsilon_2^* = -\frac{3}{5}, \quad \varepsilon_3^* = -\frac{12}{5}\varepsilon_0$$

$$u = 0, \quad v = -\frac{12}{5} L\varepsilon_0$$

Prob. 1.29 The same three-bar truss as in Prob. 1.28 except the structure is now subjected to a horizontal load Q. The load Q is first increased to the plastic limit Q_p and then is unloaded to zero. Afterward, Q is increased in the reversed direction until all three bars yield again and then is unloaded again to zero.

a. Determine the elastic limit load Q_e and the plastic limit load Q_p in the initial loading.
b. Determine the elastic limit load Q_e' and the plastic limit load Q_p' in the reversed loading.
c. Determine the residual stresses and strains of the bars and the residual horizontal and vertical displacements at the load point at the end of the loading history.

Answer:

a. $P_e = P_p = \sqrt{3}\,A\sigma_0$; b. $P_e' = P_p' = -\sqrt{3}\,A\sigma_0$

c. At the end of the loading history, no residual stresses, no residual strains, and no residual displacements.

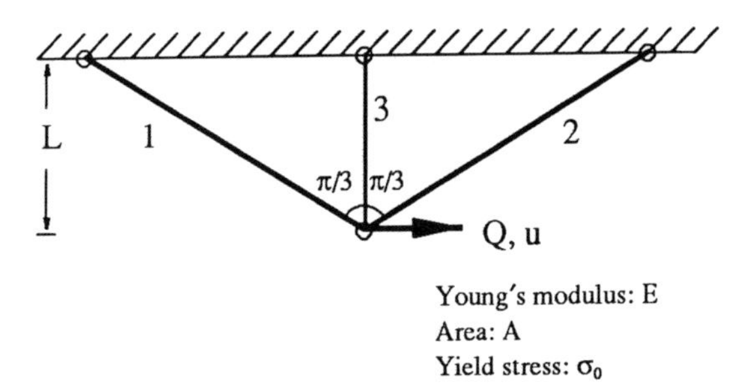

Figure P1.29.

Prob. 1.30 The same three-bar truss as in Prob. 1.28 is now subjected to the combined horizontal load Q and the vertical load P, as shown in Fig. 1.30. For the following loading paths, determine the residual strains and displacements at the load point at the end of the loading paths:

a. Increase P first to the limit load P_p. Then, maintain the same vertical displacement (i.e., $\Delta v = 0$) while increase Q until the plastic collapse of the structure.
b. Increase Q first to the limit load Q_p. Then, maintain the same load Q (i.e., $\Delta Q = 0$) while increase the load P until the plastic collapse of the structure.
c. Increase the loads P and Q proportionally, $P : Q = 1 : \sqrt{3}$, until the plastic collapse of the structure.

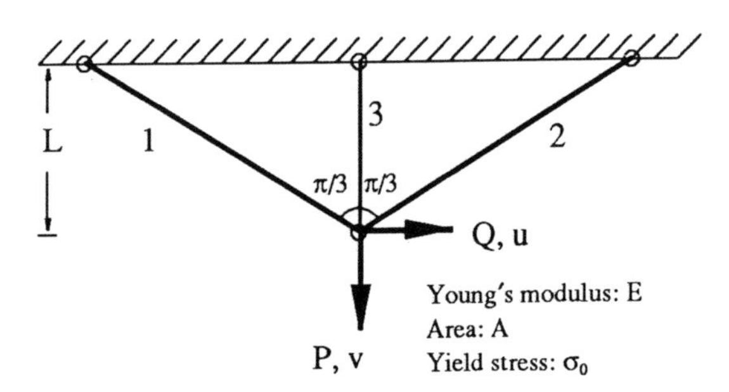

Figure P1.30.

Solution: The purpose of this problem is to show the load path dependency of elastic-plastic problems. The basic equations in the total form and in the incremental from are given below.

Equilibrium

$$\frac{1}{2}\sigma_1 - \frac{1}{2}\sigma_2 = \frac{1}{\sqrt{3}}\frac{Q}{A}, \qquad \frac{1}{2}\Delta\sigma_1 - \frac{1}{2}\Delta\sigma_2 = \frac{1}{\sqrt{3}}\frac{\Delta Q}{A}$$

$$\frac{1}{2}\sigma_1 + \frac{1}{2}\sigma_2 + \sigma_3 = \frac{P}{A}, \qquad \frac{1}{2}\Delta\sigma_1 + \frac{1}{2}\Delta\sigma_2 + \Delta\sigma_3 = \frac{\Delta P}{A}$$

Compatibility

$$\varepsilon_1 = \frac{\sqrt{3}}{4}\frac{u}{L} + \frac{1}{4}\frac{v}{L}, \qquad \Delta\varepsilon_1 = \frac{\sqrt{3}}{4}\frac{\Delta u}{L} + \frac{1}{4}\frac{\Delta v}{L}$$

$$\varepsilon_2 = -\frac{\sqrt{3}}{4}\frac{u}{L} + \frac{1}{4}\frac{v}{L}, \qquad \Delta\varepsilon_2 = -\frac{\sqrt{3}}{4}\frac{\Delta u}{L} + \frac{1}{4}\frac{\Delta v}{L}$$

$$\varepsilon_3 = \frac{v}{L}, \qquad \Delta\varepsilon_3 = \frac{\Delta v}{L}$$

and

$$\varepsilon_3 = 2(\varepsilon_1 + \varepsilon_2), \qquad \Delta\varepsilon_3 = 2(\Delta\varepsilon_1 + \Delta\varepsilon_2)$$

The elastic solution is valid for all cases and can be easily obtained as

$$\sigma_1 = \frac{1}{5}\frac{P}{A} + \frac{1}{\sqrt{3}}\frac{Q}{A}, \quad \sigma_2 = \frac{1}{5}\frac{P}{A} - \frac{1}{\sqrt{3}}\frac{Q}{A}, \quad \sigma_3 = \frac{4}{5}\frac{P}{A}$$

Loading Path a: Increase P first

From Prob. 1.28, we know the plastic limit load P_p is $2A\sigma_0$. At $P = P_p$, we have

$$\sigma_1 = \sigma_2 = \sigma_3 = \sigma_0, \quad \varepsilon_1 = \varepsilon_2 = \varepsilon_0, \quad \varepsilon_3 = 4\varepsilon_0, \qquad u = 0, \qquad \frac{v}{L} = 4\varepsilon_0$$

From the condition $\Delta v = 0$ while increasing Q, we obtain $\Delta\varepsilon_3 = 0$, and thus, $\Delta\sigma_3 = 0$. Since $\sigma_1 = \sigma_0$, increasing Q causes σ_2 to decrease. Thus, Bar 2 is in an unloading state. Thus, both σ_1 and σ_3 do not change during the process of increasing Q. From the basic equations, we obtain

$$-\frac{1}{2}\Delta\sigma_2 = \frac{1}{\sqrt{3}}\frac{\Delta Q}{A}, \qquad \frac{1}{2}\Delta\sigma_2 = \frac{\Delta P}{A}$$

Therefore, the load P must decrease as Q increases by the relationship

$$\frac{\Delta P}{A} = -\frac{1}{\sqrt{3}}\frac{\Delta Q}{A}$$

In the process of increasing Q, Bar 2 is unloaded first and then loaded in a reversed loading direction. The structure fails when Bar 2 yields in the reversed loading. Let

$$\sigma_2 = \sigma_0 + \Delta\sigma_2 = \sigma_0 - \frac{2\,\Delta Q}{\sqrt{3}\,A} = -\sigma_0$$

the structure fails at

$$\frac{\Delta Q}{A} = \sqrt{3}\,\sigma_0, \qquad \frac{\Delta P}{A} = -\sigma_0$$

or

$$\frac{P}{A} = \sigma_0, \qquad \frac{Q}{A} = \sqrt{3}\,\sigma_0$$

we determine at this state

$$\sigma_1 = \sigma_0, \quad \sigma_2 = -\sigma_0, \quad \sigma_3 = \sigma_0, \qquad \varepsilon_1 = 3\,\varepsilon_0, \quad \varepsilon_2 = -\varepsilon_0, \quad \varepsilon_3 = \varepsilon_0$$

$$\frac{u}{L} = \frac{8}{\sqrt{3}}\,\varepsilon_0, \qquad \frac{v}{L} = 4\,\varepsilon_0$$

Loading Path b: Increase Q first

From Prob. 1.29, we know the plastic limit load Q_p is $\sqrt{3}\,A\,\sigma_0$. At $Q = Q_p$, we have

$$\sigma_1 = \sigma_0, \quad \sigma_2 = -\sigma_0, \quad \sigma_3 = 0, \qquad \varepsilon_1 = \varepsilon_0, \quad \varepsilon_2 = -\varepsilon_0, \quad \varepsilon_3 = 0$$

$$\frac{u}{L} = \frac{4}{\sqrt{3}}\,\varepsilon_0, \qquad v = 0$$

In the process of increasing P with $\Delta Q = 0$, the equilibrium equations are

$$\frac{1}{2}\,\Delta\sigma_1 - \frac{1}{2}\,\Delta\sigma_2 = 0, \qquad \frac{1}{2}\,\Delta\sigma_1 + \frac{1}{2}\,\Delta\sigma_2 + \Delta\sigma_3 = \frac{\Delta P}{A}$$

from which we obtain $\Delta\sigma_1 = \Delta\sigma_2$. Since the only possible changes of the two stresses are $\Delta\sigma_1 \le 0$ and $\Delta\sigma_2 \ge 0$, we conclude that in the loading process of increasing P, we must have $\Delta\sigma_1 = \Delta\sigma_2 = 0$. Thus, the equilibrium equations become

$$\Delta\sigma_3 = \frac{\Delta P}{A}$$

The structure fails at

$$\frac{\Delta P}{A} = \sigma_0$$

or

$$\frac{P}{A} = \sigma_0, \qquad \frac{Q}{A} = \sqrt{3}\,\sigma_0$$

The strain in Bar 3 is obtained by the condition $\varepsilon_3 = \varepsilon_0$, while the vertical displacement is by $\frac{v}{L} = \varepsilon_0$. To determine the strains in Bar 1 and Bar 2, we note that since no unloading occurs in both Bar 1 and Bar 2, we have $\Delta\varepsilon_1 \geq 0$ and $\Delta\varepsilon_2 \leq 0$. If $\Delta\varepsilon_2 < 0$, the structure will not meet the compatibility condition at the load point, thus, we conclude $\Delta\varepsilon_2 = 0$. From this condition, we obtain

$$\frac{\Delta u}{L} = \frac{1}{\sqrt{3}}\frac{\Delta v}{L} = \frac{1}{\sqrt{3}}\varepsilon_0$$

The structure fails at

$$\sigma_1 = \sigma_0, \quad \sigma_2 = -\sigma_0, \quad \sigma_3 = \sigma_0, \qquad \varepsilon_1 = \frac{3}{2}\varepsilon_0, \quad \varepsilon_2 = -\varepsilon_0, \quad \varepsilon_3 = \varepsilon_0$$

$$\frac{u}{L} = \frac{5}{\sqrt{3}}\varepsilon_0, \qquad \frac{v}{L} = \varepsilon_0$$

The proportional loading path: $P : Q = 1 : \sqrt{3}$

Utilizing the condition $Q = \sqrt{3}\,P$ and the elastic stress-strain relations, the basic equations become

$$\frac{1}{2}\sigma_1 - \frac{1}{2}\sigma_2 = \frac{P}{A}, \qquad \frac{1}{2}\sigma_1 + \frac{1}{2}\sigma_2 = \frac{1}{5}\frac{P}{A}$$

and

$$\sigma_3 = 2(\sigma_1 + \sigma_2)$$

The elastic solution is then obtained as

$$\sigma_1 = \frac{6}{5}\frac{P}{A}, \qquad \sigma_2 = -\frac{4}{5}\frac{P}{A}, \qquad \sigma_3 = \frac{4}{5}\frac{P}{A}$$

Thus, Bar 1 yields first. The elastic limit load is obtained as $\dfrac{P_e}{A} = \dfrac{5}{6}\sigma_0$. At this load

$$\sigma_1 = \sigma_0, \qquad \sigma_2 = -\frac{2}{3}\sigma_0, \qquad \sigma_3 = \frac{2}{3}\sigma_0$$

$$\varepsilon_1 = \varepsilon_0, \qquad \varepsilon_2 = -\frac{2}{3}\varepsilon_0, \qquad \varepsilon_3 = \frac{2}{3}\varepsilon_0$$

$$\frac{u}{L} = \frac{10}{3\sqrt{3}}\varepsilon_0, \qquad \frac{v}{L} = \frac{2}{3}\varepsilon_0$$

For $P > P_e$, $\Delta\sigma_1 = 0$, we obtain

$$\Delta\sigma_2 = -2\frac{\Delta P}{A}, \qquad \Delta\sigma_3 = 2\frac{\Delta P}{A}$$

The structure fails when both Bar 2 and Bar 3 reach the yield stress. Thus, we find the plastic limit loads as

$$\frac{P}{A} = \sigma_0, \qquad \frac{Q}{A} = \sqrt{3}\,\sigma_0$$

At this state, we have

$$\sigma_1 = \sigma_0, \quad \sigma_2 = -\sigma_0, \quad \sigma_3 = \sigma_0, \qquad \varepsilon_1 = \frac{3}{2}\varepsilon_0, \quad \varepsilon_2 = -\varepsilon_0, \quad \varepsilon_3 = \varepsilon_0$$

$$\frac{u}{L} = \frac{5}{\sqrt{3}}\varepsilon_0, \qquad \frac{v}{L} = \varepsilon_0$$

Although all three load paths lead to the same final limit load state, $P = A\sigma_0$, $Q = \sqrt{3}\,A\sigma_0$, their respective strains and displacements are different from each other. This is in contrast with elastic analysis for which the final stresses and deformations in a structure are independent of the particular load path to reach its ultimate or limit state.

Prob. 1.31 The same three-bar truss and loads as in Prob. 1.30. Define a two-dimensional load space $(\overline{Q}, \overline{P})$, where $\overline{Q} = Q/A\sigma_0$ and $\overline{P} = P/A\sigma_0$. A point in this space with the coordinate $(\overline{Q}, \overline{P})$ represents a pair of load combination Q and P acting on the structure. Define: (1) the elastic limit points in the space corresponding to the elastic structure behavior; (2) the elastic-plastic points in the space corresponding to the state that at least one of the three bars yields; and (3) the plastic limit points in the space corresponding to the plastic collapse state of the structure.

An elastic limit locus in the load space is a closed curve which pass through all the elastic limit points. A plastic limit locus is another closed curve which pass through all the plastic limit points. When a loading path reaches the elastic limit locus, at least one of the three bars yields. When a loading path reaches the plastic limit locus, the structure reaches its plastic collapse state. Determine the initial elastic limit locus and the plastic limit locus of the structure.

Solution: The equilibrium equations and the elastic solution of the structure are the same as in Prob. 1.30. To determine the elastic locus, let $\sigma_1 = \pm\sigma_0$, $\sigma_2 = \pm\sigma_0$, and $\sigma_3 = \pm\sigma_0$. Utilizing the elastic solution, the elastic limit locus is found to be a polygon constructed by the following six straight lines.

$$\frac{1}{\sqrt{3}}\overline{Q} + \frac{1}{5}\overline{P} = \pm 1, \qquad -\frac{1}{\sqrt{3}}\overline{Q} + \frac{1}{5}\overline{P} = \pm 1, \qquad \frac{4}{5}\overline{P} = \pm 1$$

To determine the plastic limit locus, consider first the case of $Q \geq 0$, $P \geq 0$, i.e., the first quadrant in the load space. In this case, we have $\sigma_1 > 0$, $\sigma_3 \geq 0$ and σ_2 can be positive or negative. We have the following two possible cases of plastic limit:

a. $\sigma_1 = \sigma_3 = \sigma_0, \quad |\sigma_2| \leq \sigma_0$.
b. $\sigma_1 = \sigma_0, \sigma_2 = -\sigma_0, \quad \sigma_3 \leq \sigma_0$.

From these two cases, the plastic limit locus in the first quadrant is determined as two intersected straight lines as

$$\frac{1}{\sqrt{3}}\overline{Q}+\overline{P}=2$$

and

$$\frac{1}{\sqrt{3}}\overline{Q}=1, \qquad \overline{P}\leq 1$$

In a similar manner, the plastic limit locus in other quadrants can be determined.

Fig. S1.31 shows the elastic limit locus (thinner line) and the plastic limit locus (thicker line).

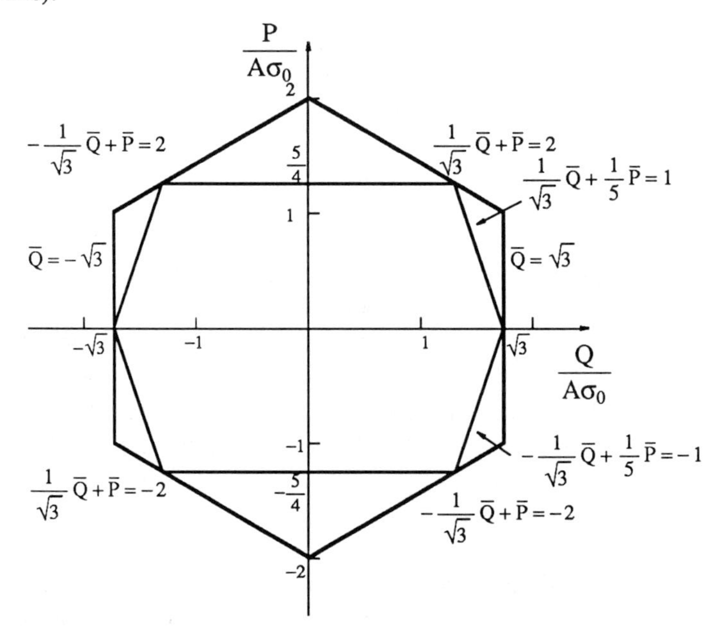

Figure S1.31. Elastic and plastic limit loci of the three-bar truss structure

Prob. 1.32 The elastic limit locus of a structure depends on the prior plastic deformation accumulated in the structure. For the same three-bar truss as in Prob. 1.31, if the load P is first increased from zero to a value $P_0 \geq P_e$, and then is unloaded to zero, verify that the closed curves shown in Fig. P1.32 are the subsequent elastic limit loci after the completion of the loading history corresponding to P_0 = 1.375, 1.5, 1.75, and 2.0. Also shown is the initial elastic limit locus corresponding to $P_0 = P_e =$ 1.25.

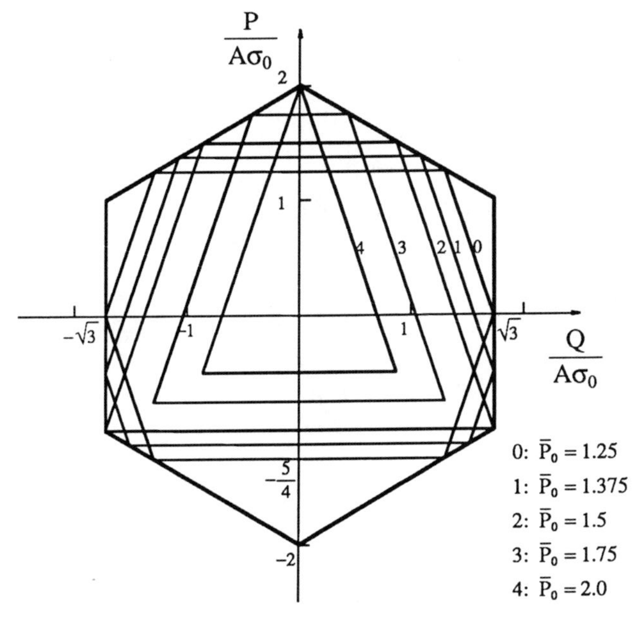

Figure P1.32. Subsequent elastic limit loci of the three-bar truss structure

Answer: The expressions for the subsequent elastic limit loci are listed below.

For P_0 *= 1.375*

$$\frac{1}{\sqrt{3}}\bar{Q}+\frac{1}{5}\bar{P}=\frac{9}{10},\quad \frac{1}{\sqrt{3}}\bar{Q}+\frac{1}{5}\bar{P}=-\frac{11}{10}$$

$$-\frac{1}{\sqrt{3}}\bar{Q}+\frac{1}{5}\bar{P}=\frac{9}{10},\quad -\frac{1}{\sqrt{3}}\bar{Q}+\frac{1}{5}\bar{P}=-\frac{11}{10}$$

$$\frac{4}{5}\bar{P}=-\frac{9}{10},\quad \frac{4}{5}\bar{P}=\frac{11}{10}$$

For P_0 *= 1.5*

$$\frac{1}{\sqrt{3}}\bar{Q}+\frac{1}{5}\bar{P}=\frac{4}{5},\quad -\frac{1}{\sqrt{3}}\bar{Q}+\frac{1}{5}\bar{P}=\frac{4}{5},\quad \frac{4}{5}\bar{P}=-\frac{4}{5},\quad \frac{4}{5}\bar{P}=\frac{6}{5}$$

For P_0 *= 1.75*

$$\frac{1}{\sqrt{3}}\bar{Q}+\frac{1}{5}\bar{P}=\frac{3}{5},\quad -\frac{1}{\sqrt{3}}\bar{Q}+\frac{1}{5}\bar{P}=\frac{3}{5},\quad \frac{4}{5}\bar{P}=-\frac{3}{5},\quad \frac{4}{5}\bar{P}=\frac{7}{5}$$

For P_0 *= 2.0*

$$\frac{1}{\sqrt{3}}\bar{Q}+\frac{1}{5}\bar{P}=\frac{2}{5}, \quad -\frac{1}{\sqrt{3}}\bar{Q}+\frac{1}{5}\bar{P}=\frac{2}{5}, \quad \frac{4}{5}\bar{P}=-\frac{2}{5}$$

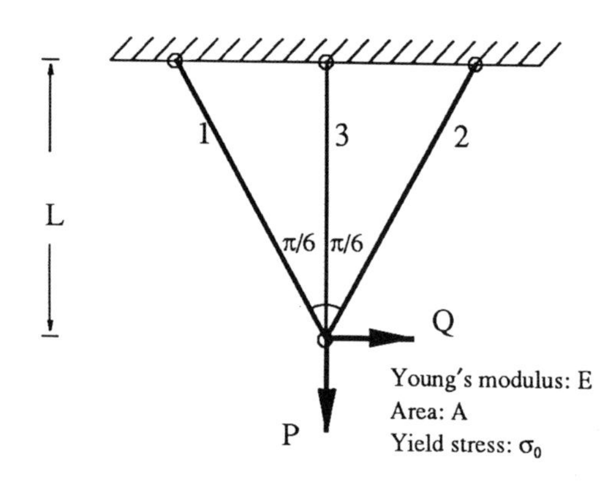

Figure P1.33.

Prob. 1.33 Determine the initial elastic limit locus and the plastic limit locus in the $(\bar{Q}, \bar{P})$ space for the three-bar truss structure shown in Fig. P1.33.

Answer: The initial elastic limit locus and the plastic limit locus are plotted in Fig. S1.33.

Prob. 1.34 A simple truss structure is subjected to a vertical force P shown in Fig. P1.34. The three bars are made of an elastic-perfectly plastic material and have the same area A, Young's modulus E, and yield stress σ_0. The load P is increased first until plastic flow occurs in the entire structure and then is decreased to zero. Afterward, the load P is increased in the reversed direction.

a. Determine the elastic and plastic limit loads P_e and P_p during the initial loading;
b. Determine the plastic limit load P_p' during the reversed loading;
c. Determine the plastic strain in each of the three bars when the load P reaches the value P_p' in the reversed loading.

Answer:

a. $P_e = \frac{3}{2}A\sigma_0, P = 2A\sigma_0$; b. $P_p' = -2A\sigma_0$;

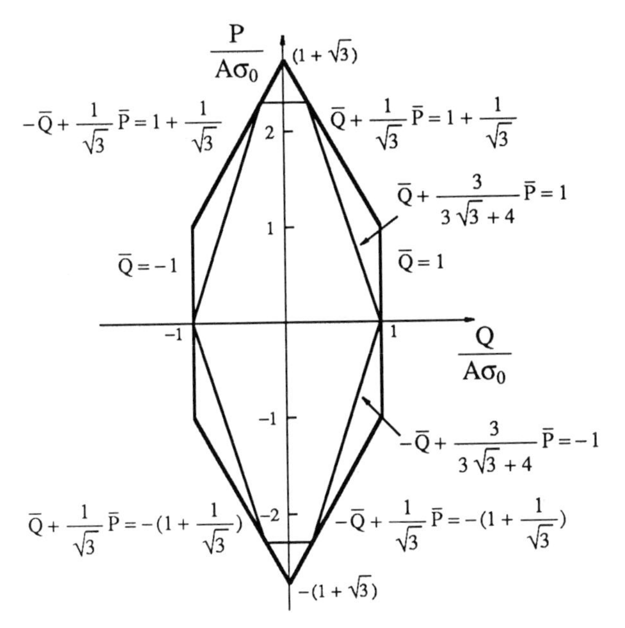

Figure S1.33. Elastic and plastic limit loci of the three-bar truss structure in Prob. 1.33

c. $\varepsilon_1^p = \varepsilon_2^p = 0,\ \varepsilon_3^p = \varepsilon_0$

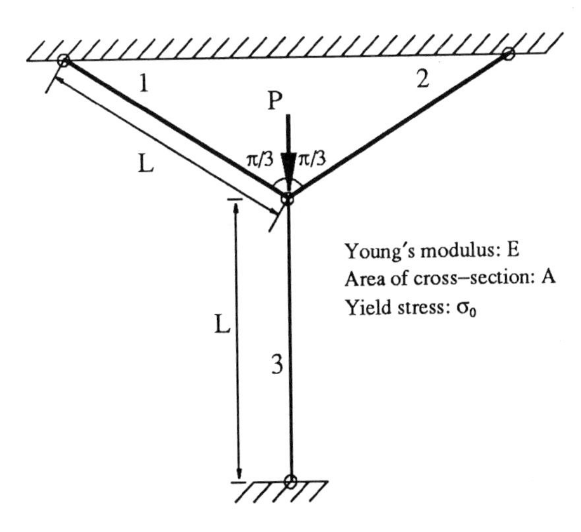

Figure P1.34.

Chapter 2

One-Dimensional Stress-Strain Analysis Software

An one-dimensional stress-strain analysis software, PLASTIC1, is introduced in this chapter. The purpose of PLASTIC1 is to help the user to understand the basic concepts of plasticity such as the hardening rules and the hardening parameters (or plastic internal variables).

For a given hardening rule and the stress-strain curve from a monotonic simple load test, PLASTIC1 computes either the stress history corresponding to a given strain history or the strain history corresponding to a given stress history.

The hardening rules implemented in PLASTIC1 include the isotropic hardening, kinematic hardening, mixed hardening, independent hardening, and perfectly plasticity.

The hardening parameters can be chosen from one of following: the equivalent plastic strain, plastic work, or plastic strain. The plastic strain is not a proper hardening parameter for developing work-hardening rules because it is reversible in a reversed loading and can not be accumulated. The reason to include the plastic strain here as a hardening parameter is to show just this particular point.

The stress-strain curve obtained from a monotonic load test is represented by one of the following three models: (1) linear-exponential; (2) Ramberg-Osgood; and (3) piecewise linear.

The formulation used in PLASTIC1 is listed in Section 2.1. Section 2.2 discusses the algorithms employed in PLASTIC1. The implementation of PLASTIC1 is described in Section 2.3. Section 2.4 is the User's Manual of this program. Several problems are given in Sections 2.5 and 2.6 to demonstrate the use of the program. A complete listing of the source code of PLASTIC1 is given in Section 2.7.

2.1 Formulation of PLASTIC1

The basic theory of one-dimensional elastic-plastic analysis has been reviewed in Chapter 1. This section lists all basic equations required for the development of PLASTIC1.

2.1.1 Basic Relations

The general incremental one-dimensional stress-strain relation can be written as

$$d\varepsilon = d\varepsilon^e + d\varepsilon^p \tag{2.1}$$

$$d\sigma = E_t\ d\varepsilon = E\ d\varepsilon^e = E_p\ d\varepsilon^p \tag{2.2}$$

$$E_t = \frac{d\sigma}{d\varepsilon}, \quad E_p = \frac{d\sigma}{d\varepsilon^p} \tag{2.3}$$

where E_t and E_p are the tangential modulus and plastic modulus respectively. The relations among these moduli are

$$\frac{1}{E_t} = \frac{1}{E} + \frac{1}{E_p} \tag{2.4}$$

or

$$E_t = \frac{E\,E_p}{E + E_p}, \quad E_p = \frac{E\,E_t}{E - E_t} \tag{2.5}$$

2.1.2 Hardening Parameters

The plastic modulus are assumed to be a function of the hardening parameter κ,

$$E_p = E_p(\kappa) \tag{2.6}$$

The hardening parameter κ can be chosen from one of the following three definitions,

$$\kappa = \varepsilon_p = \int \sqrt{d\varepsilon^p\ d\varepsilon^p} \qquad \text{equivalent plastic strain} \tag{2.7a}$$

$$\kappa = W_p = \int \sigma\ d\varepsilon^p \qquad \text{plastic work} \tag{2.7b}$$

$$\kappa = \varepsilon^p = \int d\varepsilon^p \qquad \text{plastic strain} \tag{2.7c}$$

2.1.3 Hardening Rules

2.1.3.1 The Mixed Hardening Rule

For the case of mixed hardening rule, the plastic strain increment is decomposed into two parts, $d\varepsilon^{pk}$ and $d\varepsilon^{pi}$,

$$d\varepsilon^{p} = d\varepsilon^{pk} + d\varepsilon^{pi} \tag{2.8}$$

where $d\varepsilon^{pi}$ represents the isotropic hardening part, and $d\varepsilon^{pk}$ represents the kinematic hardening part of the plastic deformation. Assuming that the ratio between $d\varepsilon^{pi}$ and $d\varepsilon^{pk}$ has a fixed value for a particular mixed hardening material satisfying

$$d\varepsilon^{pi} = M\, d\varepsilon^{p}, \quad d\varepsilon^{pk} = (1 - M)\, d\varepsilon^{p} \tag{2.9}$$

with $0 \leq M \leq 1$, the mixed hardening rule can generally be expressed as

$$|\, \sigma - c(\varepsilon^{pk})\, | = \sigma_y(\kappa^i) \tag{2.10}$$

where κ^i is the hardening parameter relating to ε^{pi}; $c(\varepsilon^{pk})$ specifies the center of the elastic range satisfying $c(0) = 0$; and $\sigma_y(\kappa^i)$ specifies the size of the elastic range satisfying $\sigma_y(0) = \sigma_0$, and σ_0 is the initial yield stress.

For simplicity, we define

$$c(\varepsilon^{pk}) = \int E_p d\varepsilon^{pk} = (1 - M)\int E_p d\varepsilon^{p} \tag{2.11}$$

and rewrite Eq.(2.10) in the form

$$|\, \sigma - c(\varepsilon^{pk})\, | = \sigma_0 + M\, [\sigma_y(\kappa) - \sigma_0] \tag{2.12}$$

For $M = 1$, Equation (2.12) reduces to the isotropic hardening rule

$$|\sigma| = \sigma_y(\kappa) \tag{2.13}$$

and for $M = 0$, Equation (2.12) reduces to the kinematic hardening rule

$$|\sigma - c(\varepsilon^{pk})| = \sigma_0 \tag{2.14}$$

2.1.3.2 Independent Hardening Rule

For the case of independent hardening rule, the tensile yield strength and the compressive yield strength are assumed to be independent of each other. The independent hardening rule can be written as

$$\sigma = \sigma_t(\kappa_t), \qquad \text{for } \sigma > 0 \tag{2.15a}$$

$$\sigma = \sigma_c(\kappa_c), \qquad \text{for } \sigma < 0 \tag{2.15b}$$

where κ_t and κ_c represent the accumulated plastic deformation in tension and in compression, respectively.

2.1.4 Experimental Stress-Strain Relationships

In PLASTIC1, the following three idealized models are used to describe the experimental stress-strain relationships in monotonic loading.

The Linear-Exponential Hardening Model

$$\sigma = E\varepsilon, \quad \text{for } \sigma < \sigma_0 \tag{2.16a}$$

$$\sigma = k\varepsilon^n, \quad \text{for } \sigma \geq \sigma_0 \tag{2.16b}$$

where k and n are constants.

The Ramberg-Osgood Model

$$\varepsilon = \frac{\sigma}{E}, \qquad \text{for } \sigma < \sigma_0 \tag{2.17a}$$

$$\varepsilon = \frac{\sigma}{E} + a\left(\frac{\sigma - \sigma_0}{b}\right)^n, \quad \text{for } \sigma \geq \sigma_0 \tag{2.17b}$$

where a, b, and n are constants. Note that this model has been modified in PLASTIC1 such that the initial yield stress is non-zero.

The Piecewise Linear Model

$$\sigma = \sigma_{i-1} + \frac{\sigma_i - \sigma_{i-1}}{\varepsilon_i - \varepsilon_{i-1}}\,\varepsilon\,, \quad \text{for } \varepsilon_{i-1} < \varepsilon < \varepsilon_i \tag{2.18}$$

$$i = 1, 2, \ldots.$$

For a given model and a particular type of hardening parameter selected, we may or may not be able to find an analytical expression for the relationship $E_p = E_p(\kappa)$. However, a tabular relation relating E_p and κ is always possible and is therefore created in PLASTIC1.

2.2 Algorithms

At a given state of current stress, strain and plastic deformation, we shall first determine the loading state corresponding to a given stress increment or a strain increment, i.e., to find out whether the given increment constitutes a plastic loading, an elastic loading or an unloading. If the increment constitutes an elastic loading or an unloading, the elastic stress-strain relationship is used to compute the new stress and strain state at the end of the increment. In the case of plastic loading, the *Euler forward* method or the *Runge-Kutta* method is used to integrate the constitutive relation to obtain the strain increment or stress increment using Eqs.(2.1) to (2.5). The hardening parameters and the current strength parameters such as $\sigma_y(\kappa)$ and $c(\kappa)$ are then updated.

In the following, the algorithms for calculating the stress increment corresponding to a given strain increment at a known stress, strain and plastic deformation state will be described. Similarly, algorithms for calculating the strain increment corresponding to a given stress increment at a known state can also be obtained with little difficulty.

2.2.1 Determining the Loading State

When the material enters the plastic state from an elastic state in an increment, a scaling factor r is determined such that at stress $\sigma + r\,\Delta\sigma$ the material begins to yield. The procedure for determining the loading state is described as follows.

1. Determine the current elastic range bounded by σ_t and σ_c:

   For isotropic hardening rule

$$\sigma_t = \sigma_y \,, \qquad \sigma_c = -\,\sigma_y \tag{2.19a}$$

   For kinematic hardening rule

$$\sigma_t = c + \sigma_0 \,, \qquad \sigma_c = c - \sigma_0 \tag{2.19b}$$

   For mixed hardening rule

$$\begin{aligned} \sigma_t &= c + \sigma_0 + M\,(\sigma_y - \sigma_0) \\ \sigma_c &= c - \sigma_0 - M\,(\sigma_y - \sigma_0) \end{aligned} \tag{2.19c}$$

   For independent hardening rule

$$\sigma_t = \sigma_t(\kappa_t) \,, \qquad \sigma_c = \sigma_c(\kappa_c) \tag{2.19d}$$

2. If $\sigma_c \le \sigma + E\,\Delta\varepsilon \le \sigma_t$, the material is currently in an elastic loading or unloading state; otherwise the material is in a plastic loading state.

3. If the material is in an elastic state at the end of the previous step and is now entering into a plastic loading state, the scaling factor r shall be determined. The strain increment is then divided into two parts: one part constitutes an elastic loading, and the remaining part constitutes a plastic loading.

2.2.2 Stress Integration

The stress integration procedure is described here. For a given strain increment constituting a plastic loading, the constitutive relations (2.1) to (2.5) will be integrated to obtain the corresponding stress increment. With little changes, this procedure can be used directly for computing the strain increment corresponding to a stress increment.

To obtain an accurate integration, the strain increment usually needs to be further divided into several sub-increments, $\Delta\tilde{\varepsilon}$,

$$\Delta\tilde{\varepsilon} = \frac{1-r}{N}\,\Delta\varepsilon \tag{2.20}$$

where N is the maximum number of sub-increment.

For each strain sub-increment, the plastic strain sub-increment is computed first,

$$\Delta\tilde{\varepsilon}^p = \frac{E_t}{E_p}\,\Delta\tilde{\varepsilon} = P(\kappa)\Delta\tilde{\varepsilon} \tag{2.21}$$

then, we obtain the stress sub-increment

$$\Delta\tilde{\sigma} = E\,(\Delta\tilde{\varepsilon} - \Delta\tilde{\varepsilon}^p) \tag{2.22}$$

Denoting

$$P_i = P(\kappa + r_i{*}\Delta\bar{\kappa}_i) \tag{2.23}$$

where $\Delta\bar{\kappa}_i$ is an increment of the hardening parameter corresponding to $\Delta\bar{\varepsilon}_i^p$ defined in the forthcoming, three integration algorithms to compute $\Delta\tilde{\varepsilon}^p$ are expressed as follows:

The Euler forward method

$$\Delta\tilde{\varepsilon}^p = P_1\,\Delta\tilde{\varepsilon} \tag{2.24a}$$

$$r_1 = 0 \tag{2.24b}$$

The second-order Runge-Kutta method

$$\Delta\tilde{\varepsilon}^p = w_1\Delta\bar{\varepsilon}_1^p + w_2\Delta\bar{\varepsilon}_2^p \tag{2.25a}$$

$$\Delta\bar{\varepsilon}_i^p = P_i\,\Delta\tilde{\varepsilon} \tag{2.25b}$$

$$r_1 = 0,\quad r_2 = 1,\qquad w_1 = w_2 = \frac{1}{2} \tag{2.25c}$$

The fourth-order Runge-Kutta method

$$\Delta\tilde{\varepsilon}^{p} = w_1\Delta\bar{\varepsilon}_1^{p} + w_2\Delta\bar{\varepsilon}_2^{p} + w_3\Delta\bar{\varepsilon}_3^{p} + w_4\Delta\bar{\varepsilon}_4^{p} \tag{2.26a}$$

$$\Delta\bar{\varepsilon}_i^{p} = P_i\,\Delta\tilde{\varepsilon} \tag{2.26b}$$

$$r_1 = 0,\ r_2 = r_3 = \frac{1}{2},\ r_4 = 1,$$

$$w_1 = w_4 = \frac{1}{6},\ w_2 = w_3 = \frac{1}{3} \tag{2.26c}$$

2.3 Implementation of PLASTIC1

The program PLASTIC1 is written primarily in C language. However, languages *Yacc* (Yet Another Compiler-Compiler) and *Lex* (A Lexical Analyzer Generator), which are available in most UNIX systems, are also used to implement the command parser of PLASTIC1.

The source code of PLASTIC1 consists of: (1) a header file, *plastic1.h*; (2) six C language files, *main.c, compute.c, modulus.c, print.c, misc.c,* and *token.c*; (3) a Yacc language file, *command.y*; and (4) a Lex language file, *lex.l*.

The file *plastic1.h* declares global variables and defines common constants. This file is included in almost all other files. The files *command.y, lex.l* and *token.c* implement the command parser. The file *main.c* contains the *main* function of PLASTIC1. The files *compute.c* and *modulus.c* perform the computation. The other two files *print.c* and *misc.c* contain functions for displaying information. In the following, the implementation of idealized stress-strain relation models and the implementation of the stress integration procedures are briefly described. A complete listing of the source code of PLASTIC1 is given in Section 2.7.

2.3.1 Implementation of Idealized Stress-Strain Models

For a given value of the hardening parameter κ, the plastic modulus E_p and yield stresses σ_y can be determined from an idealized stress-strain model. For fast processing and to treat different idealized models identically during stress or strain computation, a tabular relation relating κ with E_p and σ_y at discrete points is established in PLASTIC1. This relation can be expressed as (κ_i, E_{pi}, σ_{yi}, $i = 0, 1, 2, 3, ...$). For a selected idealized model, this tabular relation can be constructed according to the specific parameters of the model and the specific type of the hardening parameter.

This relation is stored in three double arrays: *sbk* for κ_i, *sbEp* for E_{pi}, and *sbSigmay* for σ_{yi}. These arrays all have the same dimension MAX_POINTS, which is a pre-defined constant, and are defined in the file *modulus.c*. Three functions in the file *modulus.c* are responsible for construction of the tabular relation, and each of which corresponds to a specific idealized model. The function *Ramberg_Osgood* is listed below for the *Ramberg-Osgood* model.

```
/*
 * Ramberg-Osgood Mode
 */
static void
Ramberg_Osgood()
{
    static int    sbi[] = { 10,  20,  30,  40,  70,  0 };
    static double sbf[] = {5.0, 5.0, 2.0, 2.0, 2.0, 0.0};
    auto double *pSigma = sbTestStress,
                *pEpsilon = sbTestStrain;
    auto double pl_strain = 0.0, pl_work = 0.0;
    auto double *pk = sbk, *pEp = sbEp, *pSigmay = sbSigmay;
    auto double fac1, fac2, p1, p2;
    auto int i, j, n;

    test_curve_type = TT_STRAIN;
    fac1 = ro_b / ro_n / pow(ro_a, (double) (1.0 / ro_n));
    fac2 = ro_n * fac1;
    p1 = 1.0 / ro_n - 1.0;
    p2 = p1 + 1.0;
    pSigma[0] = 0.0;
    pSigma[1] = sigmay0;
    pEpsilon[0] = 0.0;
    pEpsilon[1] = sigmay0 / E;
    pk[0] = 0.0;
    pSigmay[0] = sigmay0;
    pEp[0] = fac1 * pow((double) (pl_str_inc/500.0), p1);
    pl_str_inc /= 100.0;
    for (i = 1; i < MAX_POINTS - 1; ++i) {
        for (j = 0; sbi[j]; ++j)
            if (i == sbi[j])
                pl_str_inc *= sbf[j];
        pl_strain += pl_str_inc;
        pEp[i] = fac1 * pow(pl_strain, p1);
        pSigmay[i] = pSigma[i + 1] = sigmay0 +
                         fac2 * pow(pl_strain, p2);
        pEpsilon[i + 1] = pSigma[i + 1] / E + pl_strain;
        if (HP_PWORK == hardening_para_type) {
            pl_work += pSigmay[i] * pl_str_inc;
            pk[i] = pl_work;
        } else
```

```
            pk[i] = pl_strain;
        }
        max_kappa = pk[MAX_POINTS - 2];
        test_data_count = MAX_POINTS;
        return;
}
```

From the *Ramberg-Osgood* expression (2.17), we relate κ, σ_y and E_p with the plastic strain, ε^p,

$$\kappa = \varepsilon^p \,, \qquad \text{or } \kappa = \int \sigma d\,\varepsilon^p$$

$$\sigma_y = \sigma_0 + b \left[\frac{\varepsilon^p}{a} \right]^{\frac{1}{n}}$$

and

$$E_p = \frac{d\sigma_y}{d\varepsilon^p} = \frac{b}{an} \left[\frac{\varepsilon^p}{a} \right]^{\frac{1}{n}-1}$$

Thus, for a given value of ε^p ranging from zero to a maximum value specified by the input data, a set of data in the tabular relation can be determined from the above equations.

The tabular relation is made based on equal-spaced plastic strain points. However, for most work-hardening materials, the plastic modulus E_p changes much faster for small plastic strain than that for the large plastic strain. Thus, the range of the plastic strain in the tabular relation is divided into several segments. In each segment, the plastic strain increment is a constant. The increment increases as the plastic strain increases varying from one segment to another.

The stress-strain relation represented by the *Ramberg-Osgood* model is also generated along with the construction of the tabular relation. This relation is stored in two arrays and pointed by two pointers *pSigma* and *pEpsilon* respectively.

Once this tabular relation is established, we are ready to find E_p and σ_y for a given κ from this table. This task is performed in the function *Modulus* as listed below.

```
void
Modulus(kappa, Et, Ep, sigmay)
double kappa;
double *Ep, *Et, *sigmay;
{
    auto int i;
```

```
    auto double *pk = sbk, *pEp = sbEp, *pSigmay = sbSigmay;
    auto double k;

    if (max_kappa <= (k = ABS(kappa))) {
        *Ep = pEp[MAX_POINTS - 1];
        *sigmay = pSigmay[MAX_POINTS - 1] +
                                        *Ep * (k - max_kappa);
    } else {
        while (*pk <= k) {
            pk++;
            pEp++;
            pSigmay++;
        }
        *Ep = *(pEp-1) + (*pEp - *(pEp-1)) *
                (k - *(pk-1)) / (*pk - *(pk-1));
        *sigmay = *(pSigmay-1) + (*pSigmay - *(pSigmay-1)) *
                (k - *(pk-1)) / (*pk - *(pk-1));
    }
    *Et = *Ep * E / (*Ep + E);
    *sigmay = sigmay0 + isotropic_ratio*(*sigmay - sigmay0);
    return;
}
```

The function *Modulus* has four parameters: *kappa* is an input parameter of this function and holds the hardening parameter. *Et, Ep* and *sigmay* are the output parameters of this function and hold the calculated tangential modulus, plastic modulus, and the yield stress respectively corresponding to a given *kappa.* Values of these variables are obtained either by an interpolation (if *kappa* is less than the maximum hardening parameter) or by an extrapolation (if *kappa* is greater than the maximum hardening parameter) of the data stored in the tabular relation.

2.3.2 Implementation of the Stress Integration

The implementation of the stress integration procedure is discussed here. If a strain increment constitutes a plastic loading, the constitutive relations (2.1) to (2.5) will be integrated to obtain the corresponding stress increment. This integration is accomplished by two functions: *pl_strain_inc_by_strain* and *compute_stress.*

The function *pl_strain_inc_by_strain* computes and returns the plastic strain increment for a given strain increment which is passed to it through a parameter. The calculation of a plastic strain increment from a strain increment using different algorithms has been given in Eqs.(2.21) to (2.26). This function is listed below.

```
static double
pl_strain_inc_by_strain(epsilon_inc)
```

```
double epsilon_inc;
{
    auto double *pR, *pW;
    auto double pl_eps_inc = 0.0, pl_eps_inc_bar;
    auto double sig_inc_bar;
    auto double dk, Ep, Et, sy;
    auto int   int_count = sbCount[integration_alg_type], i;

    pR = & sbR[sbIndex[integration_alg_type]];
    pW = & sbW[sbIndex[integration_alg_type]];
    Modulus(*pk, & Et, & Ep, & sy);
    for (i = 0; i < int_count; ++i) {
        pl_eps_inc_bar = (Ep > ZERO) ?
                 Et / Ep * epsilon_inc : epsilon_inc;
        pl_eps_inc += pW[i] * pl_eps_inc_bar;
        sig_inc_bar = E * (epsilon_inc - pl_eps_inc_bar);
        switch (hardening_para_type) {
        case HP_EPSTRAIN:
                dk = ABS(pl_eps_inc_bar);
                break;
        case HP_PWORK     :
                dk = (stress + sig_inc_bar/2.0) *
                                              pl_eps_inc_bar;
                break;
        case HP_PSTRAIN     :
            dk = pl_eps_inc_bar;
                 break;
        }
        if (HR_KINEMATIC == hardening_rule ||
            HR_MIXED     == hardening_rule)
            center += pW[i]*(1.0-isotropic_ratio)*
                                         Ep * pl_eps_inc_bar;
        if (i < int_count - 1)
            Modulus(*pk + pR[i] * dk, & Et, & Ep, & sy);
    }
    return pl_eps_inc;
}
```

The integration constants are stored in two double arrays: *sbR* for the constants r_i, and *sbW* for the constants W_i. In this function, the variable for the tangential modulus, *Et*, the variable for the plastic modulus, *Ep*, and the variable for the yield stress, *sy*, are first calculated according to the current plastic deformation state. These variables are then updated during the calculation of $\Delta\bar{\varepsilon}_i$.

The function *compute_stress* computes the stress increment corresponding to a given strain increment. This function is listed below.

```
static int
compute_stress(epsilon_inc)
double epsilon_inc;
```

```
{
    auto double eps_sub_inc, pl_eps_sub_inc, sig_sub_inc;
    auto double stress_bar, elastic_ratio, sy;
    auto int    sub_inc_count, i;
    auto double Et, Ep;

    if (ABS(epsilon_inc) < ZERO * strain) {
        DisplayMessage("Stress computation",
                       "%s", "strain increment is zero!");
        return FAILED;
    }
    stress_bar = stress + E * epsilon_inc;
    /*
     * check if at the end of the step, the state is elastic
     */
    if (stress_bar >= yield_stress(-1) &&
        stress_bar <= yield_stress( 1)) {
        stress = stress_bar;
        strain += epsilon_inc;
        state  = ELASTIC;
        return SUCCESS;
    }
    /*
     * determine the elastic ratio
     */
    if (ELASTIC == state)
        elastic_ratio = (yield_stress(SIGN(epsilon_inc)) -
                        stress) / (E * epsilon_inc);
    else
        elastic_ratio = 0.0;
    state = PLASTIC;
    stress += elastic_ratio * epsilon_inc * E;
    strain += elastic_ratio * epsilon_inc;
    if (HR_INDEPENDENT == hardening_rule)
        pk = (epsilon_inc > 0.0) ? & kappat : & kappac;
    /*
     * stress integration
     */
    sub_inc_count = (1.0 - elastic_ratio) * max_sub_inc;
    sub_inc_count = MAX(sub_inc_count, 1);
    eps_sub_inc = (1.0 - elastic_ratio) * epsilon_inc /
                                          sub_inc_count;
    total_sub_inc_count += sub_inc_count;
    for (i = 0; i < sub_inc_count; ++i) {
        /* compute plastic strain sub-increment      */
        pl_eps_sub_inc=pl_strain_inc_by_strain(eps_sub_inc);
        sig_sub_inc   =E * (eps_sub_inc - pl_eps_sub_inc);
        /* update the stress and strain state       */
        strain += eps_sub_inc;
        stress += sig_sub_inc;
        /* update the plastic internal variable      */
        switch (hardening_para_type) {
```

```
        case HP_EPSTRAIN:
                *pk += ABS(pl_eps_sub_inc);
                break;
        case HP_PWORK    :
                *pk += (stress - sig_sub_inc/2.0) *
                                        pl_eps_sub_inc;
                break;
        case HP_PSTRAIN    :
            *pk += pl_eps_sub_inc;
                break;
        }
        Modulus(*pk, & Et, & Ep, & sigmay);
        if (HR_INDEPENDENT == hardening_rule)
            if (stress > 0.0)
                sigmayt = sigmay;
            else
                sigmayc = sigmay;
        pl_strain += pl_eps_sub_inc;
        pl_eq_strain += ABS(pl_eps_sub_inc);
        pl_work += (stress - sig_sub_inc / 2.0) *
                                        pl_eps_sub_inc;
        sy = yield_stress(SIGN(epsilon_inc));
        if (ABS(stress - sy) > 0.01*sy) {
            stress = sy;
            modified = TRUE;
            modify_count++;
        }
    }
    return SUCCESS;
}
```

The function performs its task in the following steps:

1. A stress increment stored in *stress_bar* corresponding to the given strain increment stored in *epsilon_inc* is first calculated based on the elastic relation. Together with the current stress, this increment is used to check the loading state. If the strain increment does not constitute a plastic loading, the stress and strain are updated based on the elastic relation, and this function returns.
2. If the strain increment constitutes a plastic loading, the scaling factor r is determined. Then the part of the strain increment which constitutes the plastic loading is further divided into several sub-increments.
3. For each strain sub-increment, the function *pl_strain_inc_by_strain* is called to calculate the plastic strain sub-increment. The corresponding stress increment is then obtained. The stress, strain, and the hardening parameter are updated.

4. According to the updated hardening parameter, the current yield stress is found and compared with the stress state. If the stress does not satisfy exactly the yield condition, $\sigma = \sigma_y$, the stress is forced to be σ_y to avoid possible error accumulation.

2.4 User's Manual of PLASTIC1

PLASTIC1 is a command-driven program. The user communicates with the program by typing in commands. A command consists of English words, and the abbreviation of these words is allowed. A command must be entered in one line. A semi-colon, ";" or a new line character (Carriage Return) indicates the end of a command.

There are two groups of commands. Commands in the first group are used for specifying parameters for a particular calculation. Commands in the second group are used for input/output, starting computation, and quitting the program.

Commands for specifying parameters consist of two parts: the name of a parameter and the value of the parameter. The name of a parameter can contain one or more words. The value of a parameter can be a number or a keyword. The name and value of a parameter in a command are separated by a colon, ":". The commands in the other group do not contain a colon. However, numbers or keywords may be included in those commands.

PLASTIC1 does not distinguish upper-case and lower-case characters. However, in the description of the commands, keywords are shown in upper-case characters. A keyword may contain one or more English words. If more than one word is contained in a keyword, these words are connected by dashes, "-". Numbers can be either integer numbers or floating point numbers and represented by NUMBER in the description.

2.4.1 PLASTIC1 Command Description

In the description of PLASTIC1 command, the abbreviated form of an English word can be obtained by neglecting all characters enclosed in the parenthesis.

2.4.1.1 Specifying Parameters

st(ress) sc(ale) : NUMBER
str(ain) sc(ale) : NUMBER

specify the scalar factors for stress and strain. The input value of stress and strain variables will be scaled (multipled) by these two factors respectively. The output results such as stress, plastic-strain, and plastic work will also be scaled (divided) by the appropriate scalar factors. If the scalar factors other than the default values are used, these two commands should be entered before entering other commands.

Default: the default values for both factors are 1.0.

Example: stress scale : 100

hard(ening) rule : KEYWORD

specify the type of hardening rule. The KEYWORD can be either IS(OTROPIC), KI(NEMATIC), MI(XED), IN(DEPENDENT), or PE(RFECTLY-PLASTIC).

Default: ISOTROPIC

Example: hardening rule : KINEMATIC

hard(ening) para(meter) : KEYWORD

specify the type of hardening parameter. KEYWORD can be either PL(ASTIC)-EQ(UIVALENT)-STR(AIN), PL(ASTIC)-WO(RK), or PL(ASTIC)-STR(AIN).

Default: PLASTIC-EQUIVALENT-STRAIN

Example: hardening parameter : PLASTIC-WORK

te(st) cu(rve) ty(pe) : KEYWORD

specify the type of the idealized model for stress-strain relation in monotonic loading. KEYWORD can be either LI(NEAR)-EXP(ONENTIAL), RA(MBERG)-OS(GOOD), or PI(ECEWISE)-LI(NEAR).

Default: LINEAR-EXPONENTIAL

Example: test curve type : RAMBERG-OSGOOD

te(st) cu(rve) var(iable) : KEYWORD

specifies the variable type of the idealized model of stress-strain relation. KEYWORD can be STR(AIN) if the function is given as stress vs. strain; or it can be PL(ASTIC)-STR(AIN) if the function is given as stress vs. plastic strain. This command applies only to the piecewise linear model.

Default: STRAIN

Example: test curve variable : PLASTIC-STRAIN

max(imum) pl(astic)-str(ain) : NUMBER

As discussed in Section 2.1, a tabular relation for E_p and κ is created in PLASTIC1 according to a given test curve and the type of hardening parameter. In doing so, the maximum value of plastic strain expected needs to be input. This command specifies the maximum plastic strain. This value should be entered properly. A too big value leads to a rough tabular relation that causes error for the nonlinear hardening model when the hardening parameter is small. When a hardening parameter is larger than the maximum value of κ, which is related to the maximum value of the plastic strain, an extrapolation of the relation is used to determine E_p. Note that this value is multipled by the strain scalar factor.

Default: 0.001

Example: maximum plastic-strain : 0.002

is(otropic) ra(tio) : NUMBER

specifies the isotropic ratio M for the mixed hardening rule.

Default: 1.0 (or the isotropic hardening rule).

Example: isotropic ratio : 0.4

int(egration) meth(od) : KEYWORD

specifies the integration method used to calculate the plastic strain increment as discussed in Section 2.2. KEYWORD can be either EU(LER), R(UNGE)-K(UTTA)-2(ND), or R(UNGE)-K(UTTA)-4(TH).

Default: EULER

Example: integration method : RUNGE-KUTTA-4TH

max(imum) sub-inc(rement) : NUMBER

specifies the maximum number of sub-increment. As discussed in Section 2.2, the strain increment usually needs to be further divided into several sub-increments to obtain an accurate integration.

Default: 1

Example: maximum sub-increment : 4

inc(rement) size : NUMBER

specifies the size of increment of strain/stress in the computation. Note that the size entered in this command is scaled (multipled) by the strain scalar factor if the strain history is given, or by the stress scalar factor if the stress history is given.

Example: increment size : 0.25

unknown : KEYWORD

specifies the unknown of the calculation. KEYWORD can be either STRAIN for computing a strain history corresponding to a given stress history, or STRESS for computing a stress history corresponding to a given strain history.

Default: STRESS

Example: unknown : STRAIN

2.4.1.2 Starting the Computation and Quitting the Program

comp(ute)

starts computation. When this command is entered, PLASTIC1 checks whether the test curve and the strain/stress history have been entered. If any one of the two group data has not been entered, the computation can not be performed.

quit

quits from PLASTIC1.

2.4.1.3 Commands for Input and Output

inp(ut) te(st) cu(rve)

starts to input the data to define the idealized model for the stress-strain relation in a monotonic loading. An interactive dialogue will start when this command is entered. For all three models described in Section 2.1, the Young's modulus will be prompted to input first. After this, the data prompted to input are different for each model. For the Linear-Exponential model, the data prompted are the initial yield stress, σ_0, and the parameter n defined in Eq.(2.16). For the Ramberg-Osgood model, the data are the initial yield stress, σ_0 and the parameters a, b, and n as defined in Eq.(2.17). For the Piecewise Linear model, the data include the number of data points to define the relation, and the strain (or plastic strain) and stress, (ε_i , σ_i), at each data point.

Note that the data input in this dialogue will not be scaled using the scalar factors except for the Piecewise Linear model, in which the strain and stress values are entered explicitly.

inp(ut) hi(story) NUMBER

specifies the number of data to describe a stress history or a strain history and starts to input the history. Following this command, the data that describes the history should be entered.

Example: input history 3
1.0
3.0
-3.0

show

lists most of the parameters for a particular computation.

show mod(ulus)

lists the E_p vs. κ relation.

show test curve

lists the stress vs. strain/plastic strain relation of the given idealized model.

show [KEYWORD], [KEYWORD], ...

lists the history of a variable or variables specified by KEYWORDs. KEYWORD can be STR(AIN), ST(RESS), PL(ASTIC)-EQ(UIVALENT)-STR(AIN), PL(ASTIC)-WO(OR), PL(ASTIC)-STR(AIN), or HARD(ENING)-PARA(METER).

Example: show STRAIN, PLASTIC-STRAIN, STRESS

2.4.2 Execute PLASTIC1

The PLASTIC1 commands can be stored in a command file. A command file can be created by a user using an editor. To make a command file readable, comment lines can be included. A comment line must start with '/*' and end with '*/' and must be entered in one line.

To start PLASTIC1, type

```
% p1 [command-file]
```

where [command-file] is the name of a PLASTIC command file. If the command file name is missing from the command line, PLASTIC1 reads commands from the standard input, and a prompt appears to prompt input as shown below

```
% p1
(Plastic1)    hard rule MI
(Plastic1)    st sc 100
(Plastic1)    ....
```

Type 'quit' or ^D terminates PLASTIC1. PLASTIC1 writes its output to the standard output, and in a UNIX system, the output can be redirected to a file.

2.5 Problems Associated with Hardening Parameters

Prob. 2.1 The σ–ε response in a simple tension test of an elastic-plastic material is approximated by the Linear-Exponential model. The material constants are

$$E = 30 \times 10^3 \text{ ksi}, \quad \sigma_0 = 30 \text{ ksi}, \quad n = 0.25$$

a. Plot the σ vs. ε_p, σ vs. W_p, and σ vs. ε^p curves for a simple tensile loading up to $\varepsilon = 0.01$.

b. For a loading-unloading-reloading loop specified by the strain history

$$\varepsilon = 0.0 \rightarrow 0.003 \rightarrow 0.0 \rightarrow -0.0015 \rightarrow 0.0 \rightarrow 0.003$$

calculate and plot the σ vs. ε and σ vs. ε^p curves using the three types of hardening parameters

1. $\kappa = \varepsilon_p$

2. $\kappa = W_p$
3. $\kappa = \varepsilon^p$

and discuss the results.

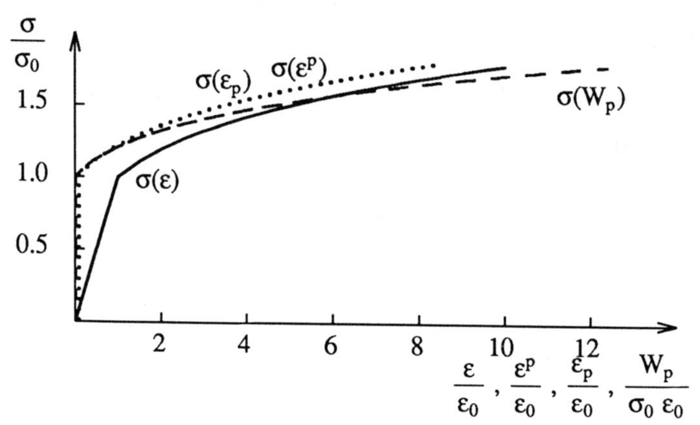

Figure S2.1a. Variation of the hardening parameters

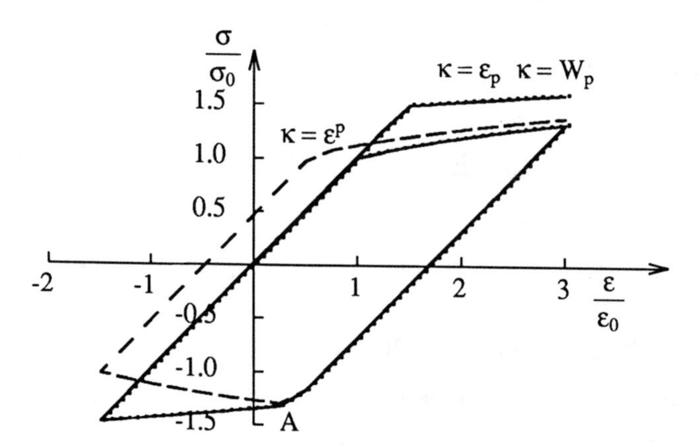

Figure S2.1b. Stress-strain curves with different hardening parameter corresponding to the following strain path: $\varepsilon = 0.0 \rightarrow 0.003 \rightarrow 0.0 \rightarrow -0.0015 \rightarrow 0.0 \rightarrow 0.003$

Solution: The command file for solving this problem is listed below.

```
/*                              */
/* command file for Prob. 2.1   */
```

```
/*                              */
stress scale: 30
strain scale: 0.001
increment size: 0.5
maximum plastic-strain: 10
test curve type: linear-exponential
integration method: r-k-4
maximum sub-increment: 4
input test curve
   30000 30 0.25
show test curve
input history 2
   0 10
compute
show
show strain, stress, plastic-strain
show stress, plastic-equivalent-strain, plastic-work
history 6
   0 3 0 -1.5 0 3
compute
show
show strain, stress, plastic-strain
quit
```

The command file listed above is only for the case of $\kappa = \varepsilon_p$. For the other two cases, the command which specifies the hardening parameter should be entered before the input of the test curve.

(a) The σ vs. ε, σ vs. ε_p, σ vs. ε^p, and σ vs. W_p curves are plotted in Fig. S2.1a. In fact, these curves except the one relating σ with ε represent the relations between the yield stress σ_y with different hardening parameters κ. From this figure, we see that the σ vs. ε^p and σ vs. ε_p curves in simple tensile loading are identical, but the σ vs. W_p curve is different from the σ vs. ε_p curve.

(b) The σ vs. ε curves for the given loading-unloading-reloading loop corresponding to three hardening parameters are plotted in Fig. S2.1b. Even though the relationships relating σ_y with ε_p and W_p are different as shown in Fig. S2.1a, the stress-strain responses obtained by using ε_p and W_p as hardening parameter respectively are very close. The reason is that both ε_p and W_p are properly defined hardening parameters, and both relations σ_y vs. ε_p and σ_y vs. W_p represent the same hardening behavior of the material.

The curve corresponding to $\kappa = \varepsilon^p$ deviates from the other two curves at point A, where the reversed plastic loading occurs, Fig. 2.1b. As noted previously, ε^p should not be used as a hardening parameter because it is reversible for a reversed loading case. In the reversed loading, ε^p decreases, leading to an increasing plastic modulus E_p and a decreasing yield stress σ_y, while for an elastic-plastic hardening material, E_p should continuously decrease while σ_y should continuously increase in a plastic loading process. This leads to a contradiction. Since the plastic modulus E_p should always be positive, the stress increment corresponding to a negative strain increment should also be negative. However, the stress should also satisfy the yield condition, $|\sigma| = \sigma_y$, and a decreasing σ_y leads to a positive stress increment in the reversed loading path when $\kappa = \varepsilon^p$ is assumed.

2.6 Problems Associated with Hardening Rules

Prob. 2.2 A Piecewise Linear model representing the simple tension test of a material is given below

$$\sigma = E\,\varepsilon \qquad\qquad 0 \le \sigma \le \sigma_0$$

$$\sigma = \sigma_0 + k\,(\varepsilon - \varepsilon_0) \qquad\qquad \sigma \ge \sigma_0,\ \ \varepsilon_0 = \frac{\sigma_0}{E}$$

where the material constants are: $E = 30 \times 10^3$ ksi, $\sigma_0 = 30$ ksi, $k = 5 \times 10^3$ ksi.

Given the following strain history

$$\varepsilon = 0.0 \rightarrow 0.004 \rightarrow -0.004 \rightarrow 0.006$$

plot the stress vs. strain and stress vs. plastic strain curves using the following hardening rules:

a. Isotropic hardening

b. Kinematic hardening

c. Independent hardening

d. Mixed hardening with M = 0.7

Solution: The command file for the mixed hardening case is listed below.

```
/*                                */
/* command file for Prob. 2.2     */
/*                                */
hardening rule: mixed
isotropic ratio: 0.7
stress scale: 30
strain scale: 0.001
increment size: 0.25
```

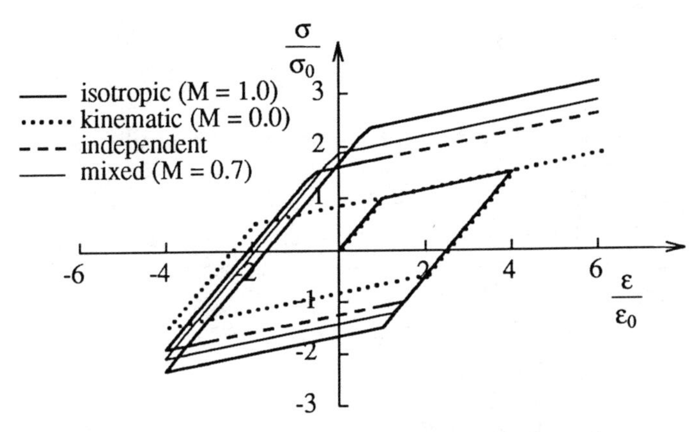

Figure S2.2a. Stress-strain curves with different hardening rules corresponding to the following strain path: $\varepsilon = 0.0 \rightarrow 0.004 \rightarrow -0.004 \rightarrow 0.006$

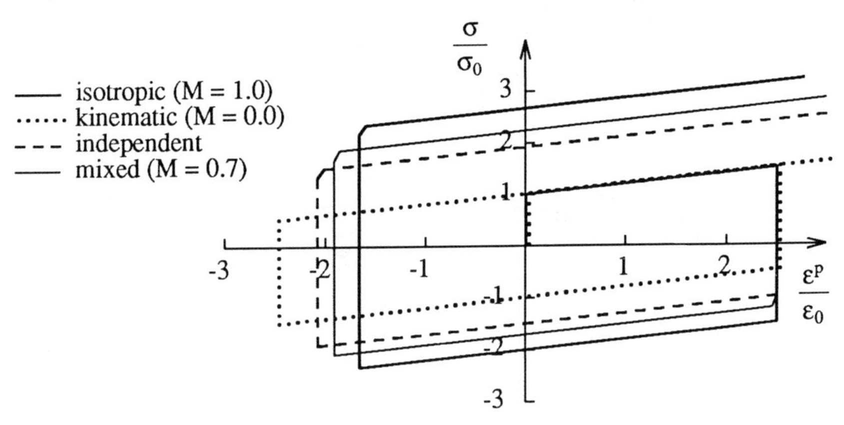

Figure S2.2b. Stress-plastic strain curves with different hardening rules corresponding to the following strain path: $\varepsilon = 0.0 \rightarrow 0.004 \rightarrow -0.004 \rightarrow 0.006$

```
maximum plastic-strain: 10
test curve type: piecewise-linear
input test curve
   30000 2 1 1 7 2
input history 5
   1.0 4.0 0.0 -4.0 6.0
compute
show test curve
```

```
show
show strain, stress, plastic-strain
quit
```

The σ vs. ε and σ vs. ε^p curves for the four types of hardening rule are plotted in Fig. S2.2a and Fig. S2.2b respectively. Note that the curves for the isotropic hardening case ($M = 1.0$) and for the kinematic hardening case ($M = 0.0$) provide upper and lower bound curves for other cases. Also note that if $M = 0.5$ is used in the mixed hardening case, the stress vs. strain, and stress vs. plastic strain curves will be identical to those of the independent hardening case. This fact will be left to the reader to explain in Prob. 2.5.

Prob. 2.3 For the material given in Problem 2.1, compute the strain history corresponding to the following stress history

$$\sigma = 0.0 \rightarrow 33.0 \rightarrow -36.0 \rightarrow 39.0 \text{ (ksi)}$$

using the four types of hardening rule as in Problem 2.2. Plot the stress vs. strain, and stress vs. plastic strain curves.

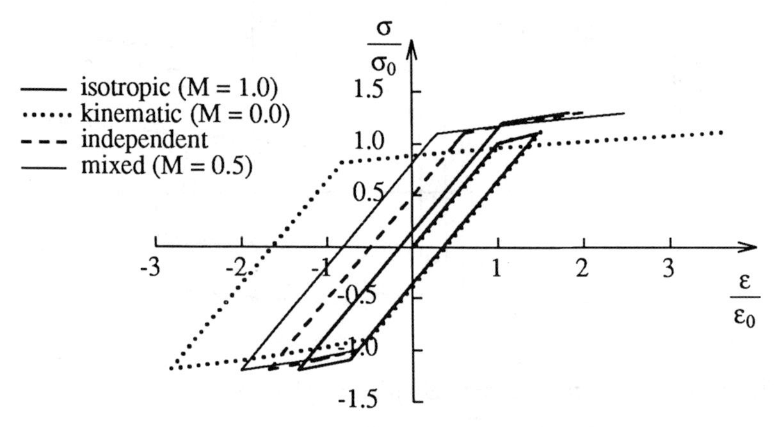

Figure S2.3a. Stress-strain curves with different hardening rules corresponding to the following stress path: $\sigma = 0.0 \rightarrow 33.0 \rightarrow -36.0 \rightarrow 39.0$

Solution: The command file for the kinematic hardening case is listed below.

```
/*                              */
/* command file for Prob. 2.3   */
/*                              */
hardening rule: kinematic
```

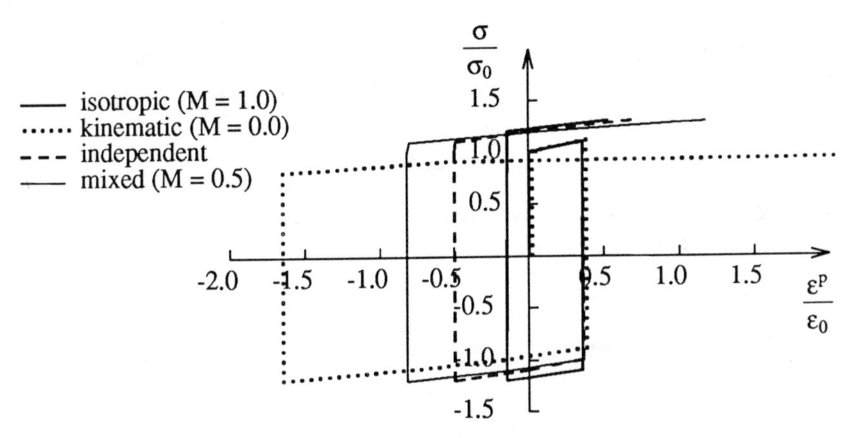

Figure S2.3b. Stress-plastic strain curve with different hardening rules corresponding to the following stress path: $\sigma = 0.0 \rightarrow 33.0 \rightarrow -36.0 \rightarrow 39.0$

```
stress scale: 30
strain scale: 0.001
maximum plastic-strain: 30
test curve type: linear-exponential
input test curve
   30000 30 0.25
show test curve
unknown: stress
increment size: 0.1
input history 7
   1 1.1 0.0 -1.1 -1.2 0 1.3
integration method: r-k-4
maximum sub-increment: 4
compute
show
show strain, stress, plastic-strain
quit
```

The σ vs. ε and σ vs. ε^p curves for the four types of hardening rule are plotted in Fig. S2.3a and Fig. S2.3b respectively. Note again that the curves for the isotropic hardening case $(M = 1.0)$ and for the kinematic hardening case $(M = 0.0)$ are the upper and lower bound curves for other cases. Since the stress history is prescribed here, more plastic deformation is expected for the kinematic hardening rule as shown in Fig. S2.3b.

Prob. 2.4 The Ramberg-Osgood model as defined in Eq.(2.17) is used to approximate the simple tension test of a material. The material constants are: $E = 30 \times 10^3$, $\sigma_0 = 30$ ksi, $a = 0.1$, $b = 30$ ksi, $n = 3$. Given the following strain history

$$\varepsilon = 0.0 \rightarrow 0.003 \rightarrow -0.004 \rightarrow 0.006$$

plot the stress vs. strain and stress vs. plastic strain curves using the following four types of hardening rule

a. Isotropic hardening

b. Kinematic hardening

c. Independent hardening

d. Mixed hardening with M = 0.3

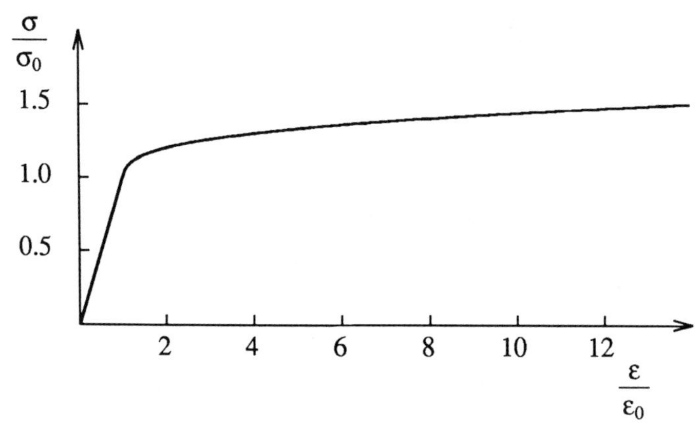

Figure S2.4a. The Ramberg-Osgood model

Solution: The command file for the independent hardening case is listed below.

```
/*                              */
/* command file for Prob. 2.4   */
/*                              */
hardening rule: independent
stress scale: 30
strain scale: 0.001
increment size: 0.25
maximum plastic-strain: 20
test curve type: ramberg-osgood
input test curve
   30000 30 0.1 30 3
show test curve
```

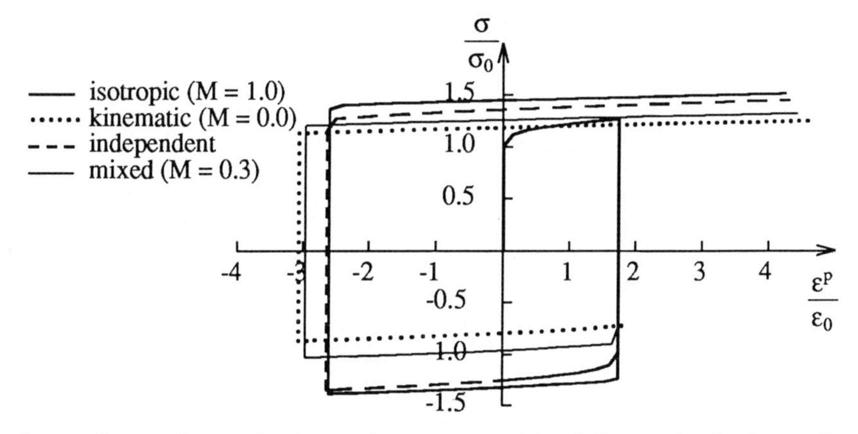

Figure S2.4b. Stress-plastic strain curves with different hardening rules corresponding to the following strain path: $\varepsilon = 0.0 \rightarrow 0.003 \rightarrow -0.004 \rightarrow 0.006$

```
input history 7
   1 3.0 0.0 -3.0 -4.0 0 6.0
integration method: r-k-4
maximum sub-increment: 10
compute
show
show strain, stress, plastic-strain
quit
```

The stress vs. strain relation for the Ramberg-Osgood model is plotted in Fig. S2.4a. The σ vs. ε^p curves for the four types of hardening rule are plotted in Fig. S2.4b.

Prob. 2.5 For the same material given in Problem 2.2, plot the stress vs. strain and stress vs. plastic strain curves using the the mixed hardening rule with M = 0.0, 0.3, 0.5, 0.7, 1.0, and the independent hardening rule. Explain why the curves obtained by the independent hardening rule is identical to those obtained by the mixed hardening rule with M = 0.5. If a non-linear hardening model such as the Ramberg-Osgood model is used, do you still expect the same conclusions ?

2.7 Source Code Listing of PLASTIC1

The following is the source code listing of PLASTIC1 in its entirety. A Makefile is also included to provide a specification on how an executable program can be built using the *make* utility of the UNIX system.

```
---------  plastic1.h  ------------------------------------------

/*
 * plastic1.h      PLASTIC1 Project
 */
#include <stdio.h>

    /*
     * logical constants
     */
#ifndef  TRUE
#define  TRUE        1
#endif
#ifndef  FALSE
#define  FALSE       0
#endif
/*
 * file pointers and line counter for the parser
 */
extern FILE *yyin, *yyout;
extern int LineNo;
/*
 * input mode
 */
extern int  input_mode;
    /*
     * constants for input modes
     */
#define STANDARD_INPUT    0
#define FILE_INPUT        1
    /*
     * returned value from the parser
     */
#define P1_CONTINUE    0
#define P1_QUIT        1
/*
 * type of computation
 */
extern int  computation_type;
    /*
     * Constants for computation type
     */
#define  CT_STRESS    0    /* given strain, compute stress */
#define  CT_STRAIN    1    /* given stress, compute strain */
```

```
/*
 * type of hardening rule
 */
extern int  hardening_rule;
    /*
     * Constants for hardening rules
     */
#define  HR_ISOTROPIC       0    /* isotropic        */
#define  HR_KINEMATIC       1    /* kinematic        */
#define  HR_MIXED           2    /* mixed            */
#define  HR_INDEPENDENT     3    /* independent      */
#define  HR_PERFECT         4    /* perfect plastic  */
/*
 * type of hardening parameter
 */
extern int  hardening_para_type;
    /*
     * Constants for hardening parameter type
     */
#define  HP_EPSTRAIN       0    /* equivalent plastic strain */
#define  HP_PWORK          1    /* plastic work              */
#define  HP_PSTRAIN        2    /* plastic strain            */
/*
 * type of the test curve
 */
extern int  test_curve_type;
    /*
     * Constants for test curve type
     */
#define  TT_STRAIN      0    /* stress-strain relation         */
#define  TT_PSTRAIN     1    /* stress-plastic strain relation */
/*
 * form of the test curve
 */
extern int  test_curve_form;
    /*
     * Constants for test curve form
     */
#define  TF_LINEXP        0    /* linear-exponential   */
#define  TF_RAMBERG       1    /* Ramberg-Osgood       */
#define  TF_PIECEWISE     2    /* piecewise linear     */
/*
 * type of integration algorithm
 */
extern int integration_alg_type;
    /*
     * Constants for integration algorithm type
     */
#define  IT_EULER       0    /* Euler method              */
#define  IT_2NDRK       1    /* 2nd order Runge-Kutta     */
#define  IT_4THRK       2    /* 4th order Runge-Kutta     */
```

```
/*
 * number of history data points
 */
extern int  history_data_count;
    /*
     * array size limits
     */
#define MAX_HISTORY    100    /* maximum history data points      */
#define MAX_POINTS     100    /* maximum test curve data points   */

extern int  test_data_count;    /* number of points in test curve  */
extern int  max_sub_inc;        /* maximum sub-increments in       */
                                /* stress/strain integration       */
extern double stress_scale;     /* stress scalar factor            */
extern double strain_scale;     /* strain scalar factor            */
extern double sigmay0;          /* initial yield stress            */
extern double isotropic_ratio;  /* isotropic ratio M               */
extern double max_pl_strain;    /* maximum plastic strain          */
extern double increment_size;   /* size of increment               */

extern double *pStress;     /* pointer to stress history           */
extern double *pStrain;     /* pointer to strain history           */
extern double *pPwork;      /* pointer to plastic work history     */
extern double *pPstrain;    /* pointer to plastic strain history   */
extern double *pPEstrain;   /* pointer to plastic equivalent       */
                            /* strain  history                     */

extern double *pTestStress;     /* stress values in the test curve */
extern double *pTestStrain;     /* strain values in the test curve */
extern double *pKappa;          /* kappa value in the Ep table     */
extern double *pModulus;        /* Ep value in the Ep table        */

extern double E;                /* Young's modulus                 */
/*
 * flags
 */
extern int test_data_input_ok;
extern int modulus_initial_ok;
extern int history_input_ok;
/*
 * functions
 */
extern void InputTestData();
extern void InputHistory();
extern void InitModulus();
extern void Modulus();
extern void Compute();
extern void DisplayMessage();
extern int  ComputeOk();
extern int  TerminalMode();
extern int  Lookup();
```

```
---------- token.c ----------------------------------------------------

/*
 * token.c         Plastic1 Project
 * command and keyword table and lookup
 */
#include <ctype.h>
#include <string.h>
#include "plastic1.h"
#include "y.tab.h"

static struct {
    char     *pKeyName;
    int      keyVal;
} KeyWords[] = {
    "comp",                    COMPUTE,
    "compute",                 COMPUTE,
    "cu",                      CURVE,
    "curve",                   CURVE,
    "eu",                      EULER,
    "euler",                   EULER,
    "hard",                    HARDENING,
    "hardening",               HARDENING,
    "hard-para",               HARDENPARA,
    "hardening-parameter",     HARDENPARA,
    "hi",                      HISTORY,
    "history",                 HISTORY,
    "in",                      INDEPENDENT,
    "independent",             INDEPENDENT,
    "inc",                     INCREMENT,
    "increment",               INCREMENT,
    "inp",                     INPUT,
    "input",                   INPUT,
    "int",                     INTEGRA,
    "integration",             INTEGRA,
    "is",                      ISOTROPIC,
    "isotropic",               ISOTROPIC,
    "ki",                      KINEMATIC,
    "kinematic",               KINEMATIC,
    "li-exp",                  LINEXP,
    "linear-exponential",      LINEXP,
    "max",                     MAXIMUM,
    "maximum",                 MAXIMUM,
    "meth",                    METHOD,
    "method",                  METHOD,
    "mi",                      MIXED,
    "mixed",                   MIXED,
    "mod",                     MODULUS,
    "modulus",                 MODULUS,
    "para",                    PARAMETER,
    "parameter",               PARAMETER,
    "pe",                      PERFECT,
```

```
    "perfect-plastic",      PERFECT,
    "pi-li",                PIECEWISE,
    "piecewise-linear",     PIECEWISE,
    "pl-eq-str",            EPLSTR,
    "plastic-equivalent-strina",    EPLSTR,
    "pl-str",               PLSTR,
    "plastic-strain",       PLSTR,
    "pl-wo",                PWORK,
    "plastic-work",         PWORK,
    "r-k-2",                RK2ND,
    "runge-kutta-2nd",      RK2ND,
    "r-k-4",                RK4TH,
    "runge-kutta-4th",      RK4TH,
    "ra",                   RATIO,
    "ratio",                RATIO,
    "ra-os",                RAMBERG,
    "ramberg-osgood",       RAMBERG,
    "ru",                   RULE,
    "rule",                 RULE,
    "sc",                   SCALE,
    "scale",                SCALE,
    "sh",                   SHOW,
    "show",                 SHOW,
    "si",                   SIZE,
    "size",                 SIZE,
    "st",                   STRESS,
    "stress",               STRESS,
    "str",                  STRAIN,
    "strain",               STRAIN,
    "sub-inc",              SUBINC,
    "sub-increment",        SUBINC,
    "te",                   TEST,
    "test",                 TEST,
    "ty",                   TYPE,
    "type",                 TYPE,
    "unk",                  UNKNOWN,
    "unknown",              UNKNOWN,
    "var",                  VARIABLE,
    "variable",             VARIABLE,
    0,                      0
};

/*
 * Lookup     look up a command or a keyword
 */
int
Lookup(pVarName)
char *pVarName;
{
    register int i;

    for (i = 0; '\000' != pVarName[i]; ++i)
```

```
        if (isupper(pVarName[i]))
            pVarName[i] = tolower(pVarName[i]);
    for (i = 0; KeyWords[i].pKeyName; i++)
        if (0 == strcmp(KeyWords[i].pKeyName, pVarName))
            return KeyWords[i].keyVal;
    return UNDEFINE;
}

----------  lex.l  --------------------------------------------------

%{
/*
 * lex.l
 * Lexical analyzer     Plastic1 Project
 */
#include <stdio.h>
#include <ctype.h>
#include "y.tab.h"
#include "plastic1.h"

%}
%%
[ \t]           {       /* skip blanks and tabs     */
                        ;
                }

[-+]?[0-9]+\.?|[-+]?[0-9]*\.[0-9]+ {     /* Number           */
                        sscanf(yytext, "%lf", &yylval.val);
                        return NUMBER;
                }

bye|quit|exit   {        /* exit          */
                        return EXIT;
                }

[a-zA-Z][a-zA-Z0-9-]*    {     /* command or keyword     */
                        return Lookup(yytext);
                }

:                {      /* just the colon : */
                        return COLON;
                }

\n|;            {     /* New line     */
                        LineNo++;
                        return NL;
                }

\/\*([^\n\"]|(\\[\n\"]))*\*\/     {     /* Comment     */
                        return COMMENT;
                }
```

```
                {     /* everything else  */
                    return yytext[0];
                }
%%

int
yywrap()
{
    return 1;
}

----------  command.y  ----------------------------------------------

%{
/*
 * command.y         Plastic1 Project
 * Parser of Plastic1 command
 */
#include <stdio.h>
#include "plastic1.h"

static double *sbVars[10];
static double  sbScales[10];
static double  scale;

%}
%union {                    /* stack type          */
    double    val;          /* number              */
    double *pV;             /* numbers             */
    int      i;             /* some thing else     */
}
%token    <val>  NUMBER
%token    <i>    COMPUTE CURVE EPLSTR HARDENING HARDENPARA HISTORY
%token    <i>    INCREMENT INDEPENDENT INPUT INTEGRA ISOTROPIC
%token    <i>    RK2ND RK4TH EULER KINEMATIC LINEXP MAXIMUM
%token    <i>    METHOD MIXED MODULUS PARAMETER PERFECT PIECEWISE
%token    <i>    PLSTR PWORK RAMBERG RATIO RULE SCALE SHOW SIZE
%token    <i>    STRAIN STRESS SUBINC TEST TYPE UNKNOWN VARIABLE
%token    <i>    EXIT COMMENT NL UNDEFINE COLON
%type     <i>    comd int_meth
%type     <i>    rule para test_form varlist
%type     <pV>   var
%%
list    :                          /*    nothing         */
                                {
                                  if (STANDARD_INPUT == input_mode)
                                      printf("(Plastic1)     ");       }
        | list NL               {
                                  if (STANDARD_INPUT == input_mode)
                                      printf("(Plastic1)     ");       }
        | list comd NL          {
                                  if (STANDARD_INPUT == input_mode)
```

```
                            printf("(Plastic1)    ");       }
        | list EXIT NL    {    /* exit          */
                            if (STANDARD_INPUT == input_mode)
                                printf("Goodbye\n");
                            return P1_QUIT;                 }
        | list COMMENT         /* does nothing        */
        | list error NL   {
                            yyerrok;
                            yyclearin;
                            if (STANDARD_INPUT == input_mode)
                                printf("(Plastic1)    ");       }
        ;
comd    : HARDENING PARAMETER COLON para
                        { hardening_para_type = $4;        }
        | HARDENING RULE COLON rule
                        { hardening_rule = $4;             }
        | INCREMENT SIZE COLON NUMBER
                        { increment_size = $4;             }
        | INTEGRA METHOD COLON int_meth
                        { integration_alg_type = $4;       }
        | ISOTROPIC RATIO COLON NUMBER
                        { isotropic_ratio = $4;            }
        | MAXIMUM PLSTR COLON NUMBER
                        { max_pl_strain = $4 * strain_scale;}
        | MAXIMUM SUBINC COLON NUMBER
                        { max_sub_inc = (int) ($4 + 0.5); }
        | STRAIN SCALE COLON NUMBER
                        { strain_scale = $4;               }
        | STRESS SCALE COLON NUMBER
                        { stress_scale = $4;               }
        | TEST CURVE TYPE COLON test_form
                        { test_curve_form = $5;            }
        | TEST CURVE VARIABLE COLON PLSTR
                        { test_curve_type = TT_PSTRAIN;    }
        | TEST CURVE VARIABLE COLON STRAIN
                        { test_curve_type = TT_STRAIN;     }
        | UNKNOWN COLON STRAIN
                        { computation_type = CT_STRAIN;    }
        | UNKNOWN COLON STRESS
                        { computation_type = CT_STRESS;    }
        | INPUT TEST CURVE
                        { InputTestData(); InitModulus(); }
        | COMPUTE
                        { if (ComputeOk()) Compute();      }
        | SHOW
                        { PrintInfo();                     }
        | SHOW MODULUS
        {
            sbVars[0] = pKappa;
            sbVars[1] = pModulus;
            sbScales[0] = (HP_PWORK == hardening_para_type) ?
                          stress_scale*strain_scale : strain_scale;
```

```
            sbScales[1] = E;
            PrintVars(2, MAX_POINTS, sbVars, sbScales);
        }
        | SHOW TEST CURVE
        {
            sbVars[0] = pTestStrain;
            sbVars[1] = pTestStress;
            sbScales[0] = strain_scale;
            sbScales[1] = stress_scale;
            PrintVars(2, test_data_count, sbVars, sbScales);
        }
        | SHOW varlist
        { PrintVars($2, history_data_count, sbVars, sbScales);}
        | INPUT HISTORY NUMBER
        {
            history_data_count = (int) ($3 + 0.5);
            InputHistory();
        }
        ;
rule    : INDEPENDENT         { $$ = HR_INDEPENDENT;     }
        | ISOTROPIC           { $$ = HR_ISOTROPIC;       }
        | KINEMATIC           { $$ = HR_KINEMATIC;       }
        | MIXED               { $$ = HR_MIXED;           }
        | PERFECT             { $$ = HR_PERFECT;         }
        ;
para    : EPLSTR              { $$ = HP_EPSTRAIN;        }
        | PLSTR               { $$ = HP_PSTRAIN;         }
        | PWORK               { $$ = HP_PWORK;           }
        ;
test_form :
          LINEXP              { $$ = TF_LINEXP;          }
        | PIECEWISE           { $$ = TF_PIECEWISE;       }
        | RAMBERG             { $$ = TF_RAMBERG;         }
        ;
var     :  STRESS
        { $$ = pStress;    scale = stress_scale;         }
        | STRAIN
        { $$ = pStrain;    scale = strain_scale;         }
        | PLSTR
        { $$ = pPstrain;   scale = strain_scale;         }
        | EPLSTR
        { $$ = pPEstrain;  scale = strain_scale;         }
        | PWORK
        { $$ = pPwork;     scale = stress_scale * strain_scale;}
        | HARDENPARA
        {
            if (HP_EPSTRAIN == hardening_para_type)
                { $$ = pPEstrain; scale = strain_scale;}
            if (HP_PSTRAIN  == hardening_para_type)
                { $$ = pPstrain; scale = strain_scale; }
            if (HP_PWORK    == hardening_para_type)
                { $$ = pPwork; scale = stress_scale*strain_scale;}
```

```
        }
        ;
varlist     : /* nothing */     { $$ = 0;                    }
        | var    { sbVars[0] = $1; sbScales[0] = scale; $$ = 1;}
        | varlist ',' var
                { sbVars[$1] = $3; sbScales[$1] = scale; $$ = $1 + 1;}
        ;
int_meth: EULER           { $$ = IT_EULER;     }
        | RK2ND           { $$ = IT_2NDRK;     }
        | RK4TH           { $$ = IT_4THRK;     }
        ;
%%

/*
 * yyerror    called for yacc syntax error
 * print warning message
 */
yyerror(s)
char *s;
{
    fprintf(stderr, "Plastic1: %s", s);
    fprintf(stderr, " near line %d\n", LineNo);
    return;
}

----------  modulus.c  ---------------------------------------------

/*
 * modulus.c    PLASTIC1 Project
 */
#include <math.h>
#include "plastic1.h"
#include "macfunct.h"

/*
 * Tabular relation of plastic modulus Ep and the yield stress
 * vs hardening parameter
 */
double sbEp[MAX_POINTS], *pModulus = sbEp;
double sbk[MAX_POINTS],  *pKappa   = sbk;
static double sbSigmay[MAX_POINTS];
/*
 * test data
 */
double sbTestStress[MAX_POINTS], *pTestStress = sbTestStress;
double sbTestStrain[MAX_POINTS], *pTestStrain = sbTestStrain;
/*
 * maximum plastic strain and hardening parameter
 */
#define MAX_STRAIN    0.001
double max_pl_strain = MAX_STRAIN;
double max_kappa;
```

```
/*
 * initial yield stress and Young's modulus
 */
double sigmay0, E;
/*
 * parameter for the Ramberg-Osgood model
 */
double ro_a, ro_b, ro_n;
/*
 * parameter for the Linear-exponential model
 */
double le_n;
/*
 * general information of the test curve
 */
int     test_curve_form = TF_LINEXP;
int     test_curve_type = TT_STRAIN;
int     test_data_count;

int     test_data_input_ok = FALSE;
int     modulus_initial_ok = FALSE;

static double pl_str_inc;

/*
 * Linear-Exponential Hardening Model
 */
static void
linear_exponential()
{
    static int    sbi[] = { 10,  20,  30,  40,  70,  0 };
    static double sbf[] = {5.0, 5.0, 2.0, 2.0, 2.0, 0.0};
    auto double *pSigma = sbTestStress, *pEpsilon = sbTestStrain;
    auto double *pk = sbk, *pEp = sbEp, *pSigmay = sbSigmay;
    auto double eps, sig, pl_eps, pl_work = 0.0, k, eps0;
    auto double fac, p, Et, eps_inc, pl_eps_inc;
    auto int i, j, n;

    test_curve_type = TT_STRAIN;
    eps0 = sigmay0 / E;
    k = E * pow(eps0, (double) (1.0 - le_n));
    fac = le_n * k;
    p = le_n - 1.0;
    pSigma[0] = 0.0;
    sig = pSigma[1] = sigmay0;
    pEpsilon[0] = 0.0;
    eps = pEpsilon[1] = eps0;
    pSigmay[0] = sigmay0;
    pk[0] = 0.0;
    Et = fac * pow(eps0, p);
    pEp[0] = E * Et / (E - Et);
```

```
    /*
     * For the linear-exponential model, we can not express the
     * stress and Ep (or Et) vs plastic strain relation explicitly.
     * Therefore, the yield stress, sigmay,  and plastic modulus,
     * Ep, vs the plastic internal variable relation are made based
     * on equal spaced strain points, instead of equal spaced plastic
     * strain points. We can make the relations based on equal-spaced
     * plastic strain points, like we did on other model, by solving
     * nonlinear equations numerically, but it is not necessary.
     */
    eps_inc = pl_str_inc / 50.0;
    for (i = 1; i < MAX_POINTS - 1; ++i) {
        for (j = 0; sbi[j]; ++j)
            if (i == sbi[j])
                eps_inc *= sbf[j];
        eps += eps_inc;
        sig  = k * pow(eps, le_n);
        Et = fac * pow(eps, p);
        pEp[i] = E * Et / (E - Et);
        pSigma[i+1] = pSigmay[i] = sig;
        pEpsilon[i+1] = eps;
        pl_eps_inc = eps - sig / E - pl_eps;
        pl_eps += pl_eps_inc;
        if (HP_PWORK == hardening_para_type) {
            pl_work += sig * pl_eps_inc;
            pk[i] = pl_work;
        } else
            pk[i] = pl_eps;
    }
    max_kappa = pk[MAX_POINTS - 2];
    test_data_count = MAX_POINTS;
    return;
}

/*
 * Ramberg-Osgood Mode
 */
static void
Ramberg_Osgood()
{
    static int    sbi[] = { 10,  20,  30,  40,  70,  0 };
    static double sbf[] = {5.0, 5.0, 2.0, 2.0, 2.0, 0.0};
    auto double *pSigma = sbTestStress, *pEpsilon = sbTestStrain;
    auto double pl_strain = 0.0, pl_work = 0.0;
    auto double *pk = sbk, *pEp = sbEp, *pSigmay = sbSigmay;
    auto double fac1, fac2, p1, p2;
    auto int i, j, n;

    test_curve_type = TT_STRAIN;
    fac1 = ro_b / ro_n / pow(ro_a, (double) (1.0 / ro_n));
    fac2 = ro_n * fac1;
    p1 = 1.0 / ro_n - 1.0;
```

```
    p2 = p1 + 1.0;
    pSigma[0] = 0.0;
    pSigma[1] = sigmay0;
    pEpsilon[0] = 0.0;
    pEpsilon[1] = sigmay0 / E;
    pk[0] = 0.0;
    pSigmay[0] = sigmay0;
    pEp[0] = fac1 * pow((double) (pl_str_inc/500.0), p1);
    pl_str_inc /= 100.0;
    for (i = 1; i < MAX_POINTS - 1; ++i) {
        for (j = 0; sbi[j]; ++j)
            if (i == sbi[j])
                pl_str_inc *= sbf[j];
        pl_strain += pl_str_inc;
        pEp[i] = fac1 * pow(pl_strain, p1);
        pSigmay[i] = pSigma[i + 1] = sigmay0 +
                            fac2 * pow(pl_strain, p2);
        pEpsilon[i + 1] = pSigma[i + 1] / E + pl_strain;
        if (HP_PWORK == hardening_para_type) {
            pl_work += pSigmay[i] * pl_str_inc;
            pk[i] = pl_work;
        } else
            pk[i] = pl_strain;
    }
    max_kappa = pk[MAX_POINTS - 2];
    test_data_count = MAX_POINTS;
    return;
}

/*
 * Piecewise Linear Model
 */
static void
piecewise_linear()
{
    auto double *pSigma = sbTestStress, *pEpsilon = sbTestStrain;
    auto double stress, pl_strain = 0.0, Ep, pl_work = 0.0;
    auto double *pk = sbk, *pEp = sbEp, *pSigmay = sbSigmay;
    auto int i, n;
    /*
     * convert the test curve to be stress vs. plastic strain
     */
    if (TT_STRAIN == test_curve_type) {
        pSigma   = (double *)malloc(sizeof(double)*test_data_count);
        pEpsilon = (double *)malloc(sizeof(double)*test_data_count);
        for (i = 1; i < test_data_count; ++i) {
            pEpsilon[i - 1] = sbTestStrain[i] - sbTestStress[i]/E;
            pSigma[i-1] = sbTestStress[i];
        }
        pEpsilon[0] = 0.0;
    }
    pSigmay[0] = sigmay0 = pSigma[0];
```

```
    stress = pSigma[0];
    Ep = (pSigma[1] - stress) / (pEpsilon[1] - pl_strain);
    pk[0] = 0.0;
    pEp[0] = Ep;
    n = 1;
    for (i = 1; i < MAX_POINTS; ++i) {
        pl_strain += pl_str_inc;
        if (pl_strain > pEpsilon[n] && n < test_data_count - 2) {
            Ep = (pSigma[n + 1] - pSigma[n]) /
                 (pEpsilon[n + 1] - pEpsilon[n]);
            stress = pSigma[n];
            ++n;
        }
        pEp[i] = Ep;
        pSigmay[i] = stress + Ep * (pl_strain - pEpsilon[n - 1]);
        if (HP_PWORK == hardening_para_type) {
            pl_work += pSigmay[i] * pl_str_inc;
            pk[i] = pl_work;
        } else
            pk[i] = pl_strain;
    }
    max_kappa = pk[MAX_POINTS - 1];
    if (TT_STRAIN == test_curve_type) {
        free((char *) pSigma);
        free((char *) pEpsilon);
    }
    return;
}

/*
 * initialize the Ep table
 */
void
InitModulus()
{
    if (! test_data_input_ok)
        return;
    pl_str_inc = max_pl_strain / MAX_POINTS;
    switch (test_curve_form) {
    case TF_LINEXP     :
        linear_exponential();
        break;
    case TF_RAMBERG     :
        Ramberg_Osgood();
        break;
    case TF_PIECEWISE     :
        piecewise_linear();
        break;
    }
    modulus_initial_ok = TRUE;
    return;
}
```

```
/*
 * input the definition of material properties
 */
void
InputTestData()
{
    if (STANDARD_INPUT == input_mode)
        printf("\nEnter the data as prompted in the follows\n");
    if (STANDARD_INPUT == input_mode)
        printf("Young's modulus:\t");
    fscanf(yyin, "%lf", & E);
    switch (test_curve_form) {
    case TF_LINEXP    :    /* linear-exponential model     */
        {
            if (STANDARD_INPUT == input_mode)
                printf("Initial yield stress:\t");
            fscanf(yyin, "%lf", & sigmay0);
            if (STANDARD_INPUT == input_mode)
                printf("The parameter n:\t");
            fscanf(yyin, "%lf", & le_n);
            break;
        }
    case TF_RAMBERG    :    /* Ramberg-Osgood model         */
        {
            if (STANDARD_INPUT == input_mode)
                printf("Initial yield stress:\t");
            fscanf(yyin, "%lf", & sigmay0);
            if (STANDARD_INPUT == input_mode)
                printf("The parameter a:\t");
            fscanf(yyin, "%lf", & ro_a);
            if (STANDARD_INPUT == input_mode)
                printf("The parameter b:\t");
            fscanf(yyin, "%lf", & ro_b);
            if (STANDARD_INPUT == input_mode)
                printf("The parameter n:\t");
            fscanf(yyin, "%lf", & ro_n);
            break;
        }
    case TF_PIECEWISE :    /* piecewise-linear model      */
        {
            auto int i;
            if (STANDARD_INPUT == input_mode)
                printf("\nNumber of test data points:\t");
            fscanf(yyin, "%d", & test_data_count);
            if (STANDARD_INPUT == input_mode)
                printf("\nEnter the data in increasing order\n");
            if (STANDARD_INPUT == input_mode)
                printf("The (0.0 0.0) will be added to the data\n");
            if (STANDARD_INPUT == input_mode)
                printf("\t(1)\t0.0 0.0\n");
            sbTestStrain[0] = sbTestStress[0] = 0.0;
            for (i = 1; i <= test_data_count; ++i) {
```

```
                if (STANDARD_INPUT == input_mode)
                    printf("\t(%d)\t", i + 1);
                fscanf(yyin, "%lf %lf",
                    & sbTestStrain[i], & sbTestStress[i]);
                sbTestStrain[i] *= strain_scale;
                sbTestStress[i] *= stress_scale;
            }
            test_data_count++;
            break;
        }
    }
    test_data_input_ok = TRUE;
}

/*
 * calculate the tangential modulus pointed by pEt
 * the plastic modulus pointed by pEp and
 * the yield stress pointed by sigmay
 * according to a given kappa using the Ep table
 */
void
Modulus(kappa, Et, Ep, sigmay)
double kappa;
double *Ep, *Et, *sigmay;
{
    auto int i;
    auto double *pk = sbk, *pEp = sbEp, *pSigmay = sbSigmay;
    auto double k;

    if (max_kappa <= (k = ABS(kappa))) {
        *Ep = pEp[MAX_POINTS - 1];
        *sigmay = pSigmay[MAX_POINTS - 1] + *Ep * (k - max_kappa);
    } else {
        while (*pk <= k) {
            pk++;
            pEp++;
            pSigmay++;
        }
        *Ep = *(pEp-1) + (*pEp - *(pEp-1)) *
                (k - *(pk-1)) / (*pk - *(pk-1));
        *sigmay = *(pSigmay-1) + (*pSigmay - *(pSigmay-1)) *
                (k - *(pk-1)) / (*pk - *(pk-1));
    }
    *Et = *Ep * E / (*Ep + E);
    *sigmay = sigmay0 + isotropic_ratio * (*sigmay - sigmay0);
    return;
}
```

---------- compute.c --

```
/*
 * compute.c    PLASTIC1 Project
 */
#include "plastic1.h"
#include "macfunct.h"

int  computation_type   = CT_STRESS;
int  hardening_rule = HR_ISOTROPIC;
int  hardening_para_type = HP_EPSTRAIN;
int  history_data_count = MAX_HISTORY;
int  integration_alg_type = IT_EULER;
int  max_sub_inc = 1;
int  history_input_ok = FALSE;

double sbStress[MAX_HISTORY], *pStress = sbStress;
double sbStrain[MAX_HISTORY], *pStrain = sbStrain;
double sbPwork[MAX_HISTORY],  *pPwork = sbPwork;
double sbPstrain[MAX_HISTORY], *pPstrain = sbPstrain;
double sbPEstrain[MAX_HISTORY], *pPEstrain = sbPEstrain;

double isotropic_ratio = 1.0;
double increment_size = 1.0;

#define SUCCESS  0
#define FAILED   1

#define ELASTIC  0
#define PLASTIC  1

#define ZERO     0.001

static int state;         /* flag of stress state */
static int error;         /* flag of error state  */
static double center;     /* elastic center       */
static double sigmayt, sigmayc, sigmay;
static double stress, strain;
static double pl_work, pl_strain, pl_eq_strain;
static double kappat, kappac, *pk;
/*
 * integration coefficients
 */
static double sbR[] = { 0.0,
            0.0, 1.0,
            0.0, 0.5, 0.5, 1.0 };
/*
 * integration weight factors
 */
static double sbW[] = { 1.0,
            0.5, 0.5,
            0.166666666667, 0.166666666667,
```

```
            0.333333333333, 0.333333333333 };
static int sbIndex[] = { 0, 1, 3 };
static int sbCount[] = { 1, 2, 4 };

static int modified, modify_count, total_sub_inc_count;

/*
 * return the yield stress in the direction of stress increasing
 */
static double
yield_stress(stress_sign)
int stress_sign;
{
    switch (hardening_rule) {
                /* In fact, the first three cases          */
                /* and the fifth case can be dealt         */
                /* with together. However, to be more      */
                /* readable, they are separated in         */
                /* individual cases.                       */

    case HR_ISOTROPIC     :
        return stress_sign * sigmay;

    case HR_KINEMATIC     :
        return (stress_sign > 0.0) ?
            (center + sigmay0) : (center - sigmay0);

    case HR_MIXED         :
        return (stress_sign > 0.0) ?
            (center + sigmay) : (center - sigmay);

    case HR_INDEPENDENT     :
        return (stress_sign > 0.0) ? sigmayt : - sigmayc;

    case HR_PERFECT         :
        return stress_sign * sigmay0;
    }
}

/*
 * compute plastic strain increment for a given strain increment
 * using either one of the Euler method, the 2nd order Runge-Kutta
 * or the 4th order Runge-Kutta method
 *
 * return the plastic strain increment
 */
static double
pl_strain_inc_by_strain(epsilon_inc)
double epsilon_inc;
{
    auto double *pR, *pW;
    auto double pl_eps_inc = 0.0, pl_eps_inc_bar, sig_inc_bar;
```

```
    auto double dk, Ep, Et, sy;
    auto int    int_count = sbCount[integration_alg_type], i;

    pR = & sbR[sbIndex[integration_alg_type]];
    pW = & sbW[sbIndex[integration_alg_type]];
    Modulus(*pk, & Et, & Ep, & sy);
    for (i = 0; i < int_count; ++i) {
        pl_eps_inc_bar = (Ep > ZERO) ?
                         Et / Ep * epsilon_inc : epsilon_inc;
        pl_eps_inc += pW[i] * pl_eps_inc_bar;
        sig_inc_bar = E * (epsilon_inc - pl_eps_inc_bar);
        switch (hardening_para_type) {
        case HP_EPSTRAIN:
                dk = ABS(pl_eps_inc_bar);
                break;
        case HP_PWORK    :
                dk = (stress + sig_inc_bar/2.0) * pl_eps_inc_bar;
                break;
        case HP_PSTRAIN    :
            dk = pl_eps_inc_bar;
                break;
        }
        if (HR_KINEMATIC == hardening_rule ||
            HR_MIXED     == hardening_rule)
            center += pW[i]*(1.0-isotropic_ratio)*Ep*pl_eps_inc_bar;
        if (i < int_count - 1)
            Modulus(*pk + pR[i] * dk, & Et, & Ep, & sy);
    }
    return pl_eps_inc;
}

/*
 * for a given strain increment, compute the stress increment
 * and update the strain, stress and plastic deformation state
 */
static int
compute_stress(epsilon_inc)
double epsilon_inc;
{
    auto double eps_sub_inc, pl_eps_sub_inc, sig_sub_inc;
    auto double stress_bar, elastic_ratio, sy;
    auto int    sub_inc_count, i;
    auto double Et, Ep;

    if (ABS(epsilon_inc) < ZERO * strain) {
        DisplayMessage("Stress computation",
                        "%s", "strain increment is zero!");
        return FAILED;
    }
    stress_bar = stress + E * epsilon_inc;
```

```
/*
 * check if at the end of the step, the state is elastic
 */
if (stress_bar >= yield_stress(-1) &&
    stress_bar <= yield_stress( 1)) {
    stress = stress_bar;
    strain += epsilon_inc;
    state  = ELASTIC;
    return SUCCESS;
}
/*
  * determine the elastic ratio
 */
if (ELASTIC == state)
    elastic_ratio = (yield_stress(SIGN(epsilon_inc)) - stress) /
                          (E * epsilon_inc);
else
    elastic_ratio = 0.0;
state = PLASTIC;
stress += elastic_ratio * epsilon_inc * E;
strain += elastic_ratio * epsilon_inc;
if (HR_INDEPENDENT == hardening_rule)
    pk = (epsilon_inc > 0.0) ? & kappat : & kappac;
/*
  * stress integration
 */
sub_inc_count = (1.0 - elastic_ratio) * max_sub_inc;
sub_inc_count = MAX(sub_inc_count, 1);
eps_sub_inc = (1.0 - elastic_ratio) * epsilon_inc /
              sub_inc_count;
total_sub_inc_count += sub_inc_count;
for (i = 0; i < sub_inc_count; ++i) {
    /* compute plastic strain sub-increment    */
    pl_eps_sub_inc = pl_strain_inc_by_strain(eps_sub_inc);
    sig_sub_inc     = E * (eps_sub_inc - pl_eps_sub_inc);
    /* update the stress and strain state    */
    strain += eps_sub_inc;
    stress += sig_sub_inc;
    /* update the plastic internal variable    */
    switch (hardening_para_type) {
    case HP_EPSTRAIN:
            *pk += ABS(pl_eps_sub_inc);
            break;
    case HP_PWORK     :
            *pk += (stress - sig_sub_inc/2.0) * pl_eps_sub_inc;
            break;
    case HP_PSTRAIN     :
        *pk += pl_eps_sub_inc;
               break;
    }
    Modulus(*pk, & Et, & Ep, & sigmay);
    if (HR_INDEPENDENT == hardening_rule)
```

```
            if (stress > 0.0)
                sigmayt = sigmay;
            else
                sigmayc = sigmay;
        pl_strain += pl_eps_sub_inc;
        pl_eq_strain += ABS(pl_eps_sub_inc);
        pl_work += (stress - sig_sub_inc / 2.0) * pl_eps_sub_inc;
        sy = yield_stress(SIGN(epsilon_inc));
        if (ABS(stress - sy) > 0.01*sy) {
            stress = sy;
            modified = TRUE;
            modify_count++;
        }
    }
    return SUCCESS;
}

/*
 * compute plastic strain increment for a given stress increment
 * using either one of the Euler method, the 2nd order Runge-Kutta
 * or the 4th order Runge-Kutta method
 * return the plastic strain increment
 */
static double
pl_strain_inc_by_stress(sigma_inc)
double sigma_inc;
{
    auto double *pR, *pW;
    auto double pl_eps_inc = 0.0, pl_eps_inc_bar;
    auto double dk, Ep, Et, sy, el_eps_inc = sigma_inc / E;
    auto int    int_count = sbCount[integration_alg_type], i;

    pR = & sbR[sbIndex[integration_alg_type]];
    pW = & sbW[sbIndex[integration_alg_type]];
    Modulus(*pk, & Et, & Ep, & sy);
    for (i = 0; i < int_count; ++i) {
        if (Et <= ZERO) {
            error = TRUE;
            DisplayMessage("Strain computation", "%s\n%s",
                           "tangential modulus is zero !"
                           "can not continue.");
            return pl_eps_inc;
        }
        pl_eps_inc_bar = Et / (Ep - Et) * el_eps_inc;
        pl_eps_inc += pW[i] * pl_eps_inc_bar;
        switch (hardening_para_type) {
        case HP_EPSTRAIN:
                dk = ABS(pl_eps_inc_bar);
                break;
        case HP_PWORK    :
                dk = (stress + sigma_inc/2.0) * pl_eps_inc_bar;
                break;
```

```
        case HP_PSTRAIN    :
            dk = pl_eps_inc_bar;
                break;
        }
        if (HR_KINEMATIC == hardening_rule ||
            HR_MIXED     == hardening_rule)
            center += pW[i]*(1.0-isotropic_ratio)*Ep*pl_eps_inc_bar;
        if (i < int_count - 1)
            Modulus(*pk + pR[i] * dk, & Et, & Ep, & sy);
    }
    return pl_eps_inc;
}

/*
 * for a given stress increment, compute the strain increment
 * and update the strain, stress and plastic deformation state
 */
static int
compute_strain(sigma_inc)
double sigma_inc;
{
    auto double eps_sub_inc, pl_eps_sub_inc, sig_sub_inc;
    auto double stress_bar, elastic_ratio, sy;
    auto int    sub_inc_count, i;
    auto double Et, Ep;

    if (ABS(sigma_inc) < ZERO) {
        DisplayMessage("Strain computation", "%s\n",
                       "stress increment is zero!");
        return FAILED;
    }
    stress_bar = stress + sigma_inc;

    /*
     * check if at the end of the step, the state is elastic
     */
    if (stress_bar >= yield_stress(-1) &&
        stress_bar <= yield_stress( 1)) {
        stress = stress_bar;
        strain += sigma_inc / E;
        state  = ELASTIC;
        return SUCCESS;
    }
    /*
     * determine the elastic ratio
     */
    if (ELASTIC == state)
        elastic_ratio = (yield_stress(SIGN(sigma_inc)) - stress) /
                                      sigma_inc;
    else
        elastic_ratio = 0.0;
    state = PLASTIC;
```

```
    stress += elastic_ratio * sigma_inc;
    strain += elastic_ratio * sigma_inc / E;
    if (HR_INDEPENDENT == hardening_rule)
        pk = (sigma_inc > 0.0) ? & kappat : & kappac;
    /*
     * strain integration
     */
    sub_inc_count = (1.0 - elastic_ratio) * max_sub_inc;
    sub_inc_count = MAX(sub_inc_count, 1);
    sig_sub_inc = (1.0 - elastic_ratio) * sigma_inc / sub_inc_count;
    for (i = 0; i < sub_inc_count; ++i) {
        /* compute plastic strain sub-increment    */
        pl_eps_sub_inc = pl_strain_inc_by_stress(sig_sub_inc);
        if (error) {
            error = FALSE;
            return FAILED;
        }
        eps_sub_inc    = sig_sub_inc / E + pl_eps_sub_inc;
        /* update the stress and strain state    */
        strain += eps_sub_inc;
        stress += sig_sub_inc;
        /* update the plastic internal variable    */
        switch (hardening_para_type) {
        case HP_EPSTRAIN:
                *pk += ABS(pl_eps_sub_inc);
                break;
        case HP_PWORK    :
                *pk += (stress - sig_sub_inc/2.0) * pl_eps_sub_inc;
                break;
        case HP_PSTRAIN    :
            *pk += pl_eps_sub_inc;
                  break;
        }
        Modulus(*pk, & Et, & Ep, & sigmay);
        if (HR_INDEPENDENT == hardening_rule)
            if (stress > 0.0)
                sigmayt = sigmay;
            else
                sigmayc = sigmay;
        pl_strain += pl_eps_sub_inc;
        pl_eq_strain += ABS(pl_eps_sub_inc);
        pl_work += (stress - sig_sub_inc / 2.0) * pl_eps_sub_inc;
        sy = yield_stress(SIGN(sigma_inc));
        if (ABS(stress - sy) > 0.01*sy)
            modified = TRUE;
    }
    return SUCCESS;
}
```

```
/*
 * compute the stress/strain history according to a given
 *              strain/stress history
 * loop over each steps in the given history
 */
void
Compute()
{
    auto int i;

    modified = FALSE;
    modify_count = 0;
    total_sub_inc_count = 0;
    if (HR_ISOTROPIC == hardening_rule ||
        HR_INDEPENDENT == hardening_rule)
        isotropic_ratio = 1.0;
    if (HR_KINEMATIC == hardening_rule)
        isotropic_ratio = 0.0;
    for (i = 0; i < MAX_HISTORY; ++i) {
        sbPwork[i] = sbPstrain[i] = sbPEstrain[i] = 0.0;
        if (CT_STRESS == computation_type)
            sbStress[i] = 0.0;
        else
            sbStrain[i] = 0.0;
    }
    stress = strain = 0.0;
    pl_strain = pl_work = pl_eq_strain = 0.0;
    kappat = kappac = 0.0;
    pk = & kappat;
    center = 0.0;
    sigmayt = sigmayc = sigmay = sigmay0;
    state  = ELASTIC;
    for (i = 1; i < history_data_count; ++i) {
        if (CT_STRESS == computation_type) {
            if (SUCCESS !=
                    compute_stress(sbStrain[i] - sbStrain[i - 1]))
            break;
            sbStress[i] = stress;
            strain = sbStrain[i];
        } else {
            if (SUCCESS !=
                    compute_strain(sbStress[i] - sbStress[i - 1]))
            break;
            sbStrain[i] = strain;
            stress = sbStress[i];
        }
        sbPwork[i] = pl_work;
        sbPstrain[i] = pl_strain;
        sbPEstrain[i] = pl_eq_strain;
    }
```

```
    if (modified) {
        if (CT_STRESS == computation_type)
            DisplayMessage("Stress computation", "%s %d %s %d %s\n",
                    "Stress has been modified in ", modify_count,
                    "sub_increments of total ", total_sub_inc_count,
                    "sub_increments to satisfy the yield condition");
        else
            DisplayMessage("Stress computation", "%s\n",
                      "The strain integration has to be refined ",
                      "to satisfy the yield condition");
    }
    return;
}

/*
 * input and interpolate a given strain or stress history
 */
void
InputHistory()
{
    auto int i, j, n, inc_count = 0;
    auto double sbHistory[MAX_HISTORY], *pHis
    auto double scale, inc_size, inc_r_size;

    if (STANDARD_INPUT == input_mode)
        printf("Enter the %s history data\n",
        (CT_STRAIN == computation_type) ? "stress" : "strain");
    if (CT_STRESS == computation_type) {
        pHis = sbStrain;
        scale = strain_scale;
    } else {
        pHis = sbStress;
        scale = stress_scale;
    }
    inc_r_size = increment_size * scale;
    for (i = 0; i < history_data_count; ++i) {
        if (STANDARD_INPUT == input_mode)
            printf("\t(%d)\t", i + 1);
        fscanf(yyin, "%lf", & sbHistory[i]);
        sbHistory[i] *= scale;
    }
    pHis[0] = 0.0;
    if (0.0 != sbHistory[0]) {
        n = (int) ABS(sbHistory[0] / inc_r_size);
        n = MAX(n, 1);
        inc_size = SIGN(sbHistory[0]) * inc_r_size;
        for (j = 0; j < n && inc_count != (MAX_HISTORY - 1); ++j)
            pHis[inc_count] = pHis[++inc_count-1] + inc_size;
        if (inc_count != MAX_HISTORY - 1 &&
            ((inc_size > 0.0 && pHis[inc_count] < sbHistory[0]) ||
             (inc_size < 0.0 && pHis[inc_count] > sbHistory[0])) &&
            ABS(pHis[inc_count] - sbHistory[0]) > inc_r_size / 2.0)
```

```
            ++inc_count;
        pHis[inc_count] = sbHistory[0];
    }
    for (i = 1; i < history_data_count; ++i) {
        n = (int) ABS((sbHistory[i]-sbHistory[i-1])/inc_r_size);
        n = MAX(n, 1);
        inc_size = SIGN(sbHistory[i] - sbHistory[i-1]) * inc_r_size;
        for (j = 0; j < n && inc_count != (MAX_HISTORY - 1); ++j)
            pHis[inc_count] = pHis[++inc_count-1] + inc_size;
        if (inc_count != MAX_HISTORY - 1 &&
            ((inc_size > 0.0 && pHis[inc_count] < sbHistory[i]) ||
             (inc_size < 0.0 && pHis[inc_count] > sbHistory[i])) &&
            ABS(pHis[inc_count] - sbHistory[i]) > inc_r_size / 2.0)
            ++inc_count;
        pHis[inc_count] = sbHistory[i];
    }
    history_data_count = inc_count + 1;
    history_input_ok = TRUE;
    return;
}

/*
 * check if the computation can be proceed
 */
int
ComputeOk()
{
    auto int return_value = TRUE;

    if (! test_data_input_ok) {
        DisplayMessage("Compute", "%s\n%s\n",
                       "The test data has not input yet.",
                       "Can not start computation");
        return_value = FALSE;
    }
    if (! modulus_initial_ok)
        return_value = FALSE;
    if (! history_input_ok) {
        DisplayMessage("Compute", "%s\n%s\n",
            "The history (strain or stress) has not input yet.",
            "Can not start computation");
        return_value = FALSE;
    }
    return return_value;
}
```

```
---------- print.c -----------------------------------------------

/*
 * print.c        Plasticl Project
 */
#include <stdio.h>
#include "plasticl.h"

double stress_scale = 1.0;
double strain_scale = 1.0;

/*
 * print the general information
 */
void
PrintInfo()
{
    static char *sbGiven[] =   {"Given strain history.",
                                "Given stress history." };
    static char *sbCompute[] = {"Compute stress history.",
                                "Compute strain history." };
    static char *sbHardRule[] = {"Isotropic",
                                 "Kinematic",
                                 "Mixed",
                                 "Independent",
                                 "Perfect-Plastic" };
    static char *sbHardPara[] = {"Equivalent plastic strain",
                                 "Plastic Work",
                                 "Plastic Strain" };
    static char *sbTestForm[] = {"Linear-Exponential",
                                 "Ramberg-Osgood",
                                 "Piecewise-linear" };
    static char *sbInteMeth[] = {"Euler",
                                 "2nd order Runge-Kutta",
                                 "4th order Runge-Kutta" };

    printf("\nGeneral Information\n\n");
    printf("\tHardening rule : %s\n", sbHardRule[hardening_rule]);
    printf("\tHardening parameter : %s\n",
                                 sbHardPara[hardening_para_type]);
    printf("\tStress scale : %f\n", stress_scale);
    printf("\tStrain scale : %f\n", strain_scale);
    printf("\tSimple relation (stress vs %s): %s\n",
        (TT_STRAIN == test_curve_type) ? "strain":"plastic-strain",
        sbTestForm[test_curve_form]);
    printf("\tMaximum plastic strain : %f\n",
                                      max_pl_strain/strain_scale);
    printf("\t%s\n\t%s\n", sbGiven[computation_type],
            sbCompute[computation_type]);
    printf("\tIntegration method : %s\n",
                                 sbInteMeth[integration_alg_type]);
    printf("\tMaximum sub-increments in integration : %d\n",
```

```
                                                    max_sub_inc);
    printf("\tNumber of history data : %d\n", history_data_count);
    printf("\tIncrement size : %f\n", increment_size);
    printf("\tIsotropic ratio : %f\n", isotropic_ratio);

    return;
}

/*
 * print a variable listing
 */
void
PrintVars(var_count, data_count, ppVars, pScales)
int var_count;
double **ppVars, *pScales;
{
    auto int i, j;

    printf("\n");
    for (i = 0; i < data_count; ++i) {
        for (j = 0; j < var_count; ++j)
            printf("%f\t", ppVars[j][i] / pScales[j]);
        printf("\n");
    }
    return;
}

---------- misc.c -------------------------------------------------

/*
 * misc.c        PLASTIC1 Project
 */
#include <stdio.h>
#include <varargs.h>
#include "macfunct.h"

/*
 * display message
 */
/*VARARGS0*/
void
DisplayMessage(va_alist)
va_dcl
{
    auto va_list args;
    auto char *fmt;

    va_start(args);
    fprintf(stderr, "Message from \"%s\" : ", va_arg(args, char *));
    fmt = va_arg(args, char *);
    (void) vfprintf(stderr, fmt, args);
    va_end(args);
```

```
    return;
}

----------  main.c  --------------------------------------------------

/*
 * main.c           Plastic1 Project
 */
#include <stdio.h>
#include "plastic1.h"

/*
 * storage for command parser
 */
FILE *yyin = (FILE *)0;
FILE *yyout = {stdout};
int LineNo;
int input_mode;

/*
 * terminal mode user interface
 */
int
TerminalMode(pFinput)
FILE *pFinput;
{
    yyin = pFinput;
    LineNo = 1;
    return yyparse();
}

/*
 * main function of Plastic1
 */
int
main(argc, argv)
int argc;
char **argv;
{
    auto FILE *pFtemp;

    /* check if there is an on-line argument        */
    if (argc > 1) {
        ++argv;
        if ((FILE *) 0 != (pFtemp = fopen(*argv, "r"))) {
            input mode = FILE_INPUT;
            if (P1_QUIT == TerminalMode(pFtemp))
                exit(0);
        } else {
            fprintf(stderr, "Plastic1 : can't open %s\n", argv);
            exit(1);
        }
```

```
    }
    input_mode = STANDARD_INPUT;
    (void) TerminalMode(stdin);
    exit(0);
}

---------- Makefile ----------------------------------------------

#
# Makefile for the PLASTIC1 Project
#
DESTDIR= /home/eadsbridge/chenw
BIN=     /home/goldengate/zhangh/bin
LIB=     /home/goldengate/zhangh/lib
INC=     /home/goldengate/zhangh/include

I=/usr/include
S=/usr/include/sys
L=${DESTDIR}/book/include

INCLUDE= -I$L -I${INC}
CDEFS=   -DDEBUG
DEBUG=   -g
CFLAGS=  ${DEBUG} ${CDEFS} ${INCLUDE}
YFLAGS=  -d

SRC=     main.c keyword.c print.c modulus.c compute.c misc.c
SRCx=    command.y lex.l repfile
GEN=     command.c lex.c
HDR=     plastic1.h
HDRg=    y.tab.h
OBJ=     main.o keyword.o print.o modulus.o compute.o misc.o \
         command.o lex.o
SOURCE=  Makefile ${HDR} ${SRC} ${SRCx}

all: ${OBJ}
    rm -f p1
    ${CC} -o p1 ${CFLAGS} ${OBJ} -lm -f68881

y.tab.h: command.c

command.c: command.y

lex.c:     lex.l repfile
    lex lex.l
    sed '/FILE/d' lex.yy.c | sed '/int yytchar/r repfile' > lex.c
    rm lex.yy.c

clean: FRC
    rm -f Makefile.bak *.o a.out core ${GEN} ${HDRg}

depend: ${HDRg} ${SRC} ${HDR} ${SRCx} FRC
```

```
	maketd -a ${CDEFS} ${INCLUDE} ${SRC} ${GEN}

print: source FRC
	cat ${SOURCE} | ind >source.out

lint: ${SRC} ${HDR} FRC
	lint -hx ${CDEFS} ${INCLUDE} ${SRC}

source: ${SOURCE}

spotless: clean
	rcsclean ${SRC} ${SRCx} ${HDRP}

tags: ${SRC} ${HDR}
	ctags -t ${SRC} ${HDR}

${SOURCE}:
	co $@

FRC:

# DO NOT DELETE THIS LINE - make depend DEPENDS ON IT
A=/home/eadsbridge/chenw/book/include
B=/home/goldengate/zhangh/include

main.o: $I/stdio.h main.c plastic1.h

keyword.o: $I/ctype.h $I/stdio.h keyword.c plastic1.h y.tab.h

print.o: $I/stdio.h plastic1.h print.c

modulus.o: $B/macfunct.h $I/stdio.h modulus.c plastic1.h

compute.o: $B/macfunct.h $I/stdio.h compute.c plastic1.h

misc.o: $B/macfunct.h $I/stdio.h misc.c

command.o: $B/macfunct.h $I/stdio.h command.c command.y plastic1.h

lex.o: $B/macfunct.h $I/ctype.h $I/stdio.h lex.c plastic1.h y.tab.h

# *** Do not add anything here - It will go away. ***


----------  repfile  -------------------------------------------
* this file contains the line to be inserted in lex.yy.c.

extern FILE *yyin, *yyout;
```

Chapter 3

Elastic Stress and Strain Analysis

This chapter summarizes the formulations for three-dimensional elastic stress and strain analysis. These formulations will be used in the chapters that follow.

3.1 Stress Tensor

The stress state at a point can be represented by a second-order symmetric tensor, σ_{ij}, in a right-handed Cartesian coordinate system *123* or *xyz*. In a matrix form, the stress tensor σ_{ij} can be expressed as

$$\sigma_{ij} = \begin{bmatrix} \sigma_{11} & \sigma_{12} & \sigma_{13} \\ \sigma_{21} & \sigma_{22} & \sigma_{23} \\ \sigma_{31} & \sigma_{32} & \sigma_{33} \end{bmatrix} = \begin{bmatrix} \sigma_{xx} & \sigma_{xy} & \sigma_{zx} \\ \sigma_{xy} & \sigma_{yy} & \sigma_{yz} \\ \sigma_{zx} & \sigma_{yz} & \sigma_{zz} \end{bmatrix} \tag{3.1}$$

where the diagonal terms are normal stress components, and off-diagonal terms are shear stress components. The stress tensor can also be written in engineering notation as

$$\sigma_{ij} = \begin{bmatrix} \sigma_{x} & \tau_{xy} & \tau_{zx} \\ \tau_{xy} & \sigma_{y} & \tau_{yz} \\ \tau_{zx} & \tau_{yz} & \sigma_{z} \end{bmatrix} \tag{3.2}$$

where σ represents a normal component, and τ represents a shear component of the stress state.

The stress state at a point in a stressed body can be represented either by the stress tensor σ_{ij} in the coordinate system x_i or by σ'_{ij} in the coordinate system x'_i. They are related by the following transformation laws

$$\sigma'_{ij} = l_{im} l_{jn} \sigma_{mn}, \quad \text{or} \quad \sigma_{ij} = l_{mi} l_{nj} \sigma'_{mn} \tag{3.3}$$

where l_{ij} is the directional cosines of the axis x'_i in the system x_i.

3.2 Principal Stresses and Stress Invariants

The stress vector acting on a plane in a stressed body depends on the unit normal, n_i, of the plane. If the plane is so oriented that the stress vector acting on the plane is in the same direction as n_i, then the plane is called a *principal plane*, its normal direction n_i is called a *principal direction*, and the magnitude of the stress vector is called a *principle stress*. Principal stresses and principal directions of a stress state are determined by the following equations

$$(\sigma_{ij} - \sigma\,\delta_{ij})\,n_j = 0 \tag{3.4}$$

where σ is the magnitude of principal stresses, and δ_{ij} is the *Kronecker delta* tensor. To find the principal directions n_i, we must have

$$|\,\sigma_{ij} - \sigma\,\delta_{ij}\,| = 0 \tag{3.5}$$

This leads to the *characteristic equation*

$$\sigma^3 - I_1\sigma^2 + I_2\sigma - I_3 = 0 \tag{3.6}$$

where

$$I_1 = \sigma_{ii} = \sigma_{11} + \sigma_{22} + \sigma_{33} \tag{3.7}$$

$$I_2 = \begin{vmatrix} \sigma_{22} & \sigma_{23} \\ \sigma_{32} & \sigma_{33} \end{vmatrix} + \begin{vmatrix} \sigma_{11} & \sigma_{13} \\ \sigma_{31} & \sigma_{33} \end{vmatrix} + \begin{vmatrix} \sigma_{11} & \sigma_{12} \\ \sigma_{21} & \sigma_{22} \end{vmatrix} \tag{3.8}$$

and

$$I_3 = \begin{vmatrix} \sigma_{11} & \sigma_{12} & \sigma_{13} \\ \sigma_{21} & \sigma_{22} & \sigma_{23} \\ \sigma_{31} & \sigma_{32} & \sigma_{33} \end{vmatrix} \tag{3.9}$$

where I_1, I_2 and I_3 are the first, second and third *invariants* of stress tensor respectively. The values of these invariants depend only on the stress state and are independent of coordinate systems.

It can be shown that because of symmetry of stress tensor, the characteristic equation has three real roots, known as the principal stresses. Moreover, if these three principal stresses have different values, three mutually perpendicular principal directions can be determined uniquely from Eq.(3.4). If these three principal directions are chosen as the Cartesian coordinate system, the stress tensor has only normal components, and all shear components are zero.

These three principal stresses are also invariants of the stress tensor. However, only three independent invariants exist for a second-order tensor. Thus, all stress invariants can be expressed in terms of three independent invariants. For example, the stress invariants I_1, I_2 and I_3 can be expressed in terms of the principal stresses in the form

$$I_1 = \sigma_1 + \sigma_2 + \sigma_3 \tag{3.10}$$

$$I_2 = \sigma_1\sigma_2 + \sigma_2\sigma_3 + \sigma_3\sigma_1 \tag{3.11}$$

and

$$I_3 = \sigma_1\sigma_2\sigma_3 \tag{3.12}$$

Stationary values of shear stresses with respect to coordinate transformation can also be determined as

$$\tau_1 = \frac{1}{2} \mid \sigma_2 - \sigma_3 \mid, \quad \tau_2 = \frac{1}{2} \mid \sigma_3 - \sigma_1 \mid, \quad \tau_3 = \frac{1}{2} \mid \sigma_1 - \sigma_2 \mid \tag{3.13}$$

These stationary shear stresses, called *principal shear stresses*, occur on the planes whose normals bisect the angle between principal planes. If the principal stresses are arranged in the order $\sigma_1 \geq \sigma_2 \geq \sigma_3$, the *maximum shear stress*, τ_{max}, is equal to $\frac{1}{2}(\sigma_1 - \sigma_3)$.

3.3 Deviatoric Stress Tensor and Its Invariants

The average value of all normal components of a stress tensor, called the *mean stress*, is defined as

$$p = \frac{1}{3}\sigma_{kk} = \frac{1}{3}(\sigma_{11} + \sigma_{22} + \sigma_{33}) \tag{3.14}$$

The mean stress state is a *hydrostatic stress state* and can be expressed by the *hydrostatic stress tensor* $p\delta_{ij}$. Subtracting the hydrostatic stress tensor from the stress tensor σ_{ij} results in the *deviator stress tensor* s_{ij},

$$s_{ij} = \sigma_{ij} - p\delta_{ij} \tag{3.15}$$

Note that $\delta_{ij} = 0$ and $s_{ij} = \sigma_{ij}$ for $i \neq j$ in Equation (3.15). The deviatoric stress tensor s_{ij} represents a pure shear stress state, and it plays an important role in the theory of plasticity.

The invariants associated with the deviatoric stress tensor can be produced in a similar manner as that of the stress tensor. The principal deviatoric stresses, denoted as s_1, s_2, and s_3, have the same principal directions as stress tensor. The characteristic equation for the determination of the principal deviatoric stresses is

$$\mid s_{ij} - s\delta_{ij} \mid = 0 \tag{3.16}$$

or

$$s^3 - J_1 s^2 - J_2 s - J_3 = 0 \tag{3.17}$$

where J_1, J_2, and J_3 are the first, second, and third invariants of the deviatoric stress tensor. These invariants may be expressed in different forms in terms of stress components and deviatoric stress components. The frequently used relationships are given below

$$J_1 = s_{ii} = s_{11} + s_{22} + s_{33} = s_1 + s_2 + s_3 = 0 \tag{3.18}$$

$$\begin{aligned} J_2 &= \frac{1}{2} s_{ij} s_{ji} = \frac{1}{2} (s_{11}^2 + s_{22}^2 + s_{33}^2 + 2 s_{12}^2 + 2 s_{23}^2 + 2 s_{31}^2) \\ &= \frac{1}{2} (s_1^2 + s_2^2 + s_3^2) \\ &= \frac{1}{6} [(\sigma_x - \sigma_y)^2 + (\sigma_y - \sigma_z)^2 + (\sigma_z - \sigma_x)^2] + \tau_{xy}^2 + \tau_{yz}^2 + \tau_{zx}^2 \\ &= \frac{1}{6} [(\sigma_1 - \sigma_2)^2 + (\sigma_2 - \sigma_3)^2 + (\sigma_3 - \sigma_1)^2] \end{aligned} \tag{3.19}$$

$$\begin{aligned} J_3 &= \frac{1}{3} s_{ij} s_{jk} s_{ki} = \begin{vmatrix} s_{11} & s_{12} & s_{13} \\ s_{21} & s_{22} & s_{23} \\ s_{31} & s_{32} & s_{33} \end{vmatrix} \\ &= \frac{1}{3} (s_1^3 + s_2^3 + s_3^3) = s_1 s_2 s_3 \end{aligned} \tag{3.20}$$

3.4 Geometric Representation of Stress States

Using the three principal stresses, σ_1, σ_2 and σ_3, as the coordinates, a three-dimensional stress space can be constructed. Thus, the stress state at a point in a stressed body can now be represented by a point in this stress space. Every point in the space having coordinates $(\sigma_1, \sigma_2, \sigma_3)$ represents a possible stress state. This representation of stress state is known as the *Haigh-Westergaard stress space*.

The decomposition of a stress state into a hydrostatic stress component, $p\delta_{ij}$, and a deviatoric stress component, s_{ij}, can be represented geometrically in this stress space. An arbitrary stress state is represented here as the stress vector **OP** starting from O(0, 0, 0) and ending at $P(\sigma_1, \sigma_2, \sigma_3)$ as shown in Fig.3.1. The vector **OP** can be decomposed into two components: **ON** and **NP**. The component **ON** is in the direction of the unit vector $(1/\sqrt{3}, 1/\sqrt{3}, 1/\sqrt{3})$, and the component

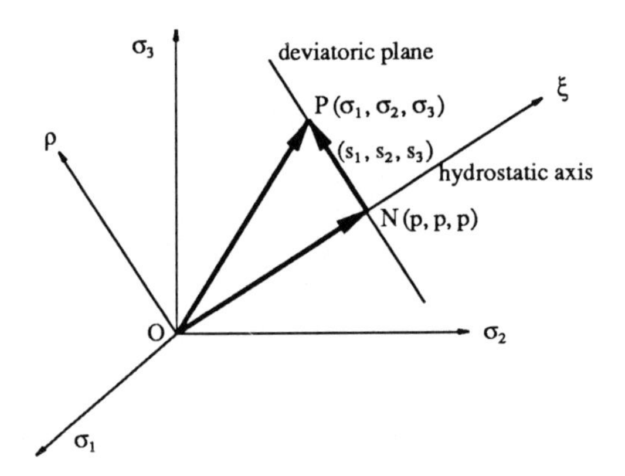

Figure 3.1. Haigh-Westergaard stress space

NP perpendicular to **ON**.

The vector **ON** represents the hydrostatic component of the stress state. The vector **ON** makes an equal angle with each of the three coordinate axes. The axis Oξ is called the *hydrostatic axis* ξ, and every point on this axis has the value $\sigma_1 = \sigma_2 = \sigma_3 = p$, or

$$\xi = \sqrt{3}\, p \tag{3.21}$$

where p is the mean stress.

The component **NP** represents the deviatoric component of the stress state, (s_1, s_2, s_3). **NP** is perpendicular to the ξ axis. Any plane perpendicular to the hydrostatic axis is called *deviatoric plane* and is expressed as

$$\frac{1}{\sqrt{3}}(\sigma_1 + \sigma_2 + \sigma_3) = \xi$$

The particular deviatoric plane that passes through the origin is called the π plane and is represented by $\xi = 0$.

Any plane containing the hydrostatic axis is called a *meridian plane.* The vector **NP** lies in a meridian plane and is perpendicular to ξ and has the ρ value

$$\rho = (s_1^2 + s_2^2 + s_3^2)^{1/2} = \sqrt{2J_2} \tag{3.22}$$

The projections of the vector **NP** and the coordinate axes σ_i on a deviatoric plane are shown in Fig.3.2. The axes σ_1', σ_2', and σ_3' are the projections of the axes σ_1, σ_2, and σ_3 respectively. The vector **N′P′** is

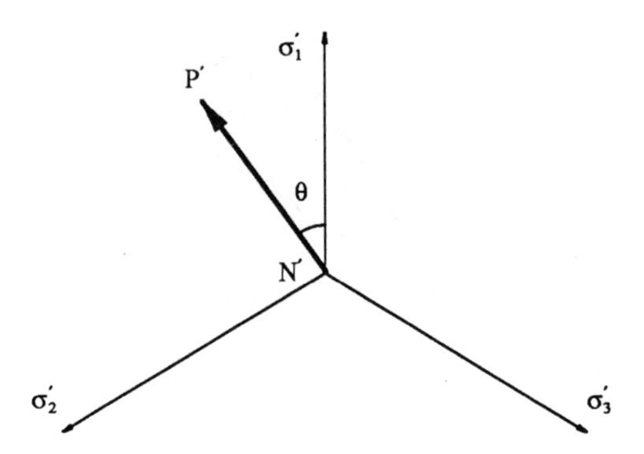

Figure 3.2. Stress state on a deviatoric plane

the projection of the vector **NP** on this plane making an angle θ with the axis σ_1'. The angle can be determined from the relation

$$\cos 3\theta = \frac{3\sqrt{3}}{2} \frac{J_3}{J_2^{3/2}} \tag{3.23}$$

The three new variables ξ, ρ, and θ as defined in Eqs.(3.21), (3.22) and (3.23) respectively can all be expressed in terms of the principal stresses through stress invariants. General stress state can therefore be expressed conveniently either in terms of $(\sigma_1, \sigma_2, \sigma_3)$ or (ξ, ρ, θ). Alternatively, all principal stresses can be expressed in terms of (ξ, ρ, θ). For example, in the range of $0 \leq \theta \leq \frac{\pi}{3}$ for $\sigma_1 \geq \sigma_2 \geq \sigma_3$, we have

$$\begin{Bmatrix} \sigma_1 \\ \sigma_2 \\ \sigma_3 \end{Bmatrix} = \begin{Bmatrix} p \\ p \\ p \end{Bmatrix} + \frac{2}{\sqrt{3}} \sqrt{J_2} \begin{Bmatrix} \cos\theta \\ \cos(\theta - 2\pi/3) \\ \cos(\theta + 2\pi/3) \end{Bmatrix}$$

$$= \frac{1}{\sqrt{3}} \begin{Bmatrix} \xi \\ \xi \\ \xi \end{Bmatrix} + \sqrt{\frac{2}{3}}\, \rho \begin{Bmatrix} \cos\theta \\ \cos(\theta - 2\pi/3) \\ \cos(\theta + 2\pi/3) \end{Bmatrix} \tag{3.24}$$

3.5 Equilibrium Equations

In tensor notation, the three equilibrium equations at a point in a stressed body are expressed as

$$\sigma_{ij,j} + F_i = 0 \tag{3.25}$$

where F_i is body force vector. In engineering notation, Equation (3.25) can be expanded to the form

$$\begin{aligned} &\frac{\partial \sigma_x}{\partial x} + \frac{\partial \tau_{xy}}{\partial y} + \frac{\partial \tau_{zx}}{\partial z} + F_x = 0 \\ &\frac{\partial \tau_{xy}}{\partial x} + \frac{\partial \sigma_y}{\partial y} + \frac{\partial \tau_{yz}}{\partial z} + F_y = 0 \\ &\frac{\partial \tau_{zx}}{\partial x} + \frac{\partial \tau_{yz}}{\partial y} + \frac{\partial \sigma_z}{\partial z} + F_z = 0 \end{aligned} \tag{3.26}$$

3.6 Strain Tensor

The strain state at a point in a deformed body can be represented by a second-order symmetric tensor. In a Cartesian coordinate system *123* or *xyz*, the strain tensor can be expressed in a matrix form as

$$\varepsilon_{ij} = \begin{bmatrix} \varepsilon_{11} & \varepsilon_{12} & \varepsilon_{13} \\ \varepsilon_{21} & \varepsilon_{22} & \varepsilon_{23} \\ \varepsilon_{31} & \varepsilon_{32} & \varepsilon_{33} \end{bmatrix} = \begin{bmatrix} \varepsilon_{xx} & \varepsilon_{xy} & \varepsilon_{zx} \\ \varepsilon_{xy} & \varepsilon_{yy} & \varepsilon_{yz} \\ \varepsilon_{zx} & \varepsilon_{yz} & \varepsilon_{zz} \end{bmatrix} \tag{3.27}$$

where the diagonal terms are the normal strain components in the directions of the coordinate axes, and the off-diagonal terms are shear strains. In engineering notation, normal strain components are expressed as ε_x, ε_y, ε_z, shear strain components are expressed as γ_{xy}, γ_{yz}, and γ_{zx}, and $\varepsilon_{xy} = \frac{1}{2}\gamma_{xy}$, etc. Thus, the strain tensor can also be expressed as

$$\varepsilon_{ij} = \begin{bmatrix} \varepsilon_x & \frac{1}{2}\gamma_{xy} & \frac{1}{2}\gamma_{zx} \\ \frac{1}{2}\gamma_{xy} & \varepsilon_y & \frac{1}{2}\gamma_{yz} \\ \frac{1}{2}\gamma_{zx} & \frac{1}{2}\gamma_{yz} & \varepsilon_z \end{bmatrix} \tag{3.28}$$

The coordinate transformation relations relating the strain tensors in different coordinate systems are

$$\varepsilon'_{ij} = l_{im} l_{jn} \varepsilon_{mn}, \quad \text{or} \quad \varepsilon_{ij} = l_{mi} l_{nj} \varepsilon'_{mn} \tag{3.29}$$

where l_{ij} is the directional cosines of the axis x'_i in the system x_i.

3.7 Principal Strains and Strain Invariants

The *principal directions* of a strain tensor, n_i, is defined as those fibers in the body in these directions that are mutually perpendicular to each other before the deformation remain mutually perpendicular to each other after the deformation. Along these directions, the shear strains are zero, and the normal strain components are called *principal strains* and denoted by ε_1, ε_2, and ε_3. The principal strains and principal directions of a strain state are determined by the following equations

$$(\varepsilon_{ij} - \varepsilon \delta_{ij}) n_j = 0 \tag{3.30}$$

This leads to the *characteristic equation*

$$\varepsilon^3 - I'_1 \varepsilon^2 + I'_2 \varepsilon - I'_3 = 0 \tag{3.31}$$

where

$$I'_1 = \varepsilon_{ii} = \varepsilon_{11} + \varepsilon_{22} + \varepsilon_{33} \tag{3.32}$$

$$I'_2 = \begin{vmatrix} \varepsilon_{22} & \varepsilon_{23} \\ \varepsilon_{32} & \varepsilon_{33} \end{vmatrix} + \begin{vmatrix} \varepsilon_{11} & \varepsilon_{13} \\ \varepsilon_{31} & \varepsilon_{33} \end{vmatrix} + \begin{vmatrix} \varepsilon_{11} & \varepsilon_{12} \\ \varepsilon_{21} & \varepsilon_{22} \end{vmatrix} \tag{3.33}$$

and

$$I'_3 = \begin{vmatrix} \varepsilon_{11} & \varepsilon_{12} & \varepsilon_{13} \\ \varepsilon_{21} & \varepsilon_{22} & \varepsilon_{23} \\ \varepsilon_{31} & \varepsilon_{32} & \varepsilon_{33} \end{vmatrix} \tag{3.34}$$

I'_1, I'_2 and I'_3 are the first, second and third *invariants* of stress tensor respectively

Here, as in the stress case, it can be shown that because of symmetry the characteristic equation has three real roots, the three principal strains. The principal strains are also invariants of strain tensor.

Similar to principal shear stresses, we can also define *principal shear strains*. In engineering notation, principal shear strains can be expressed as

$$\gamma_1 = |\, \varepsilon_2 - \varepsilon_3 \,|, \quad \gamma_2 = |\, \varepsilon_3 - \varepsilon_1 \,|, \quad \gamma_3 = |\, \varepsilon_1 - \varepsilon_2 \,| \tag{3.35}$$

3.8 Deviatoric Strain Tensor and Its Invariants

Again, as in the case of stress tensor, a strain tensor can be decomposed into two parts, a spherical part associated with a volume change and a deviatoric part associated with a shape change. This decomposition is expressed as

$$\varepsilon_{ij} = \frac{1}{3}\varepsilon_{kk}\delta_{ij} + e_{ij} \tag{3.36}$$

where $\varepsilon_{kk} = \varepsilon_v = \varepsilon_{11} + \varepsilon_{22} + \varepsilon_{33}$, called *volume strain*, represents the volumetric change per unit volume, and e_{ij} is, called the *deviatoric strain tensor*, represents a pure shear deformation or distortion.

The invariants of deviatoric strain tensor are analogous to those of deviatoric stress tensor. The characteristic equation for the determination of principal deviatoric strains, e_1, e_2, and e_3, is derived from the condition $| e_{ij} - e\delta_{ij} | = 0$ and is expressed as

$$e^3 - J_1' e^2 - J_2' e - J_3' = 0 \tag{3.37}$$

where

$$J_1' = e_{11} + e_{22} + e_{33} = e_1 + e_2 + e_3 \tag{3.38}$$

$$\begin{aligned} J_2' &= \frac{1}{2} e_{ij} e_{ij} = -(e_1 e_2 + e_2 e_3 + e_3 e_1) \\ &= \frac{1}{2}(e_{11}^2 + e_{22}^2 + e_{33}^3 + 2e_{12}^2 + 2e_{23}^2 + 2e_{31}^2) \\ &= \frac{1}{6}[(e_{11} - e_{22})^2 + (e_{22} - e_{33})^2 + (e_{33} - e_{11})^2] + \\ &\quad e_{12}^2 + e_{23}^2 + e_{31}^2 \end{aligned} \tag{3.39}$$

and

$$\begin{aligned} J_3' &= \frac{1}{3} e_{ij} e_{jk} e_{ki} = \begin{vmatrix} e_{11} & e_{12} & e_{13} \\ e_{21} & e_{22} & e_{23} \\ e_{31} & e_{32} & e_{33} \end{vmatrix} \\ &= \frac{1}{3}(e_1^2 + e_2^2 + e_3^2) = e_1 e_2 e_3 \end{aligned} \tag{3.40}$$

3.9 Compatibility Equations

For a given displacements u_i or (u, v, w) at a point in a deformed body, the corresponding strain tensor at the point can be derived from

$$\varepsilon_{ij} = \frac{1}{2}(u_{i,j} + u_{j,i}) \tag{3.41}$$

Equation (3.41) can be expanded in component form as

$$\begin{aligned}
\varepsilon_{xx} &= \frac{\partial u}{\partial x}, \quad \varepsilon_{yz} = \varepsilon_{zy} = \frac{1}{2}\left[\frac{\partial w}{\partial y} + \frac{\partial v}{\partial z}\right] \\
\varepsilon_{yy} &= \frac{\partial v}{\partial y}, \quad \varepsilon_{zx} = \varepsilon_{xz} = \frac{1}{2}\left[\frac{\partial u}{\partial z} + \frac{\partial w}{\partial x}\right] \\
\varepsilon_{zz} &= \frac{\partial w}{\partial z}, \quad \varepsilon_{xy} = \varepsilon_{yx} = \frac{1}{2}\left[\frac{\partial v}{\partial x} + \frac{\partial u}{\partial y}\right]
\end{aligned} \tag{3.42}$$

Since the six strain components are related to the three displacement components, thus, these strain components cannot be arbitrarily prescribed for a unique displacement solution. They must satisfy the following compatibility condition in order to have solution of the partial differential equations (3.42)

$$\varepsilon_{ij,kl} + \varepsilon_{kl,ij} - \varepsilon_{ik,jl} - \varepsilon_{jl,ik} = 0 \tag{3.43}$$

Since these relations will not be used in this book, we will not discuss further.

3.10 Generalized Hooke's Law

The stress-strain relations of a linear isotropic elastic material follow the generalized Hooke's law and have the following form, in the usual engineering notation

$$\begin{aligned}
\varepsilon_x &= \frac{1}{E}[\sigma_x - \nu(\sigma_y + \sigma_z)], \quad \gamma_{yz} = \frac{1}{G}\tau_{yz} \\
\varepsilon_y &= \frac{1}{E}[\sigma_y - \nu(\sigma_z + \sigma_x)], \quad \gamma_{zx} = \frac{1}{G}\tau_{zx} \\
\varepsilon_z &= \frac{1}{E}[\sigma_z - \nu(\sigma_x + \sigma_y)], \quad \gamma_{xy} = \frac{1}{G}\tau_{xy}
\end{aligned} \tag{3.44}$$

where E is the Young's modulus, ν the Poisson's ratio, and G the shear modulus. E, ν, and G are elastic material constants. The constant G is related to E and ν by

$$G = \frac{E}{2(1+\nu)} \tag{3.45}$$

Using tensor notation, Eq.(3.44) can be written in a compact form as

$$\varepsilon_{ij} = \frac{1+\nu}{E}\sigma_{ij} - \frac{\nu}{E}\sigma_{kk}\delta_{ij} \tag{3.46}$$

or expressing the stress in terms of strain, we have

$$\sigma_{ij} = \frac{E}{1+\nu}\varepsilon_{ij} + \frac{\nu E}{(1+\nu)(1-2\nu)}\varepsilon_{kk}\delta_{ij}$$

$$= 2\mu\varepsilon_{ij} + \lambda\varepsilon_{kk}\delta_{ij} \tag{3.47}$$

where λ and μ are called *Lame's elastic constants*. The relationship between the two sets of elastic constants, E, ν, G, and λ, μ are

$$\mu = G = \frac{E}{2(1+\nu)}, \quad \lambda = \frac{\nu E}{(1+\nu)(1-2\nu)} \tag{3.48}$$

The bulk modulus K defines the relationship between the mean stress or hydrostatic pressure p and its corresponding volumetric change ε_{kk} at a point

$$K = \frac{p}{\varepsilon_{kk}} \tag{3.49}$$

The constant K can also be expressed in terms of E and ν as

$$K = \frac{E}{3(1-2\nu)} \tag{3.50}$$

In fact, for an isotropic linear elastic material, there are only two independent elastic constants. Any one of those six constants described above can be expressed uniquely in terms of any two selected constants. The relationships between these constants are summarized in Table 3.1.

The stress-strain tensor Equations (3.46) and (3.47) can be written in a more compact form as

$$\varepsilon_{ij} = D_{ijkl}\sigma_{kl}, \quad \sigma_{ij} = C_{ijkl}\varepsilon_{kl} \tag{3.51}$$

D_{ijkl} is called *compliance tensor*, and C_{ijkl} *stiffness tensor*. Both D_{ijkl} and C_{ijkl} are isotropic fourth-order tensors. Comparing with Eqs.(3.46) and (3.47), we have

Table 3.1 Relationships between isotropic elastic material constants

	E	ν	G or μ	λ	K
E, ν	E	ν	$\frac{E}{2(1+\nu)}$	$\frac{\nu E}{(1+\nu)(1-2\nu)}$	$\frac{E}{3(1-2\nu)}$
G, λ	$\frac{G(3\lambda+2G)}{\lambda+G}$	$\frac{\lambda}{2(\lambda+G)}$	G	λ	$\frac{3\lambda+2G}{3}$
K, λ	$\frac{9K(K-\lambda)}{3K-\lambda}$	$\frac{\lambda}{3K-\lambda}$	$\frac{3}{2}(K-\lambda)$	λ	K
E, G	E	$\frac{E-2G}{2G}$	G	$\frac{G(E-2G)}{3G-E}$	$\frac{GE}{3(3G-E)}$
E, K	E	$\frac{3K-E}{6K}$	$\frac{3EK}{9K-E}$	$\frac{3K(3K-E)}{9K-E}$	K
K, ν	$3K(1-2\nu)$	ν	$\frac{3K(1-2\nu)}{2(1+\nu)}$	$\frac{3\nu K}{(1+\nu)}$	K
G, ν	$2G(1+\nu)$	ν	G	$\frac{2\nu G}{1-2\nu}$	$\frac{2G(1+\nu)}{3(1-2\nu)}$
λ, ν	$\frac{\lambda(1+\nu)(1-2\nu)}{\nu}$	ν	$\frac{\lambda(1-2\nu)}{2\nu}$	λ	$\frac{\lambda(1+\nu)}{3\nu}$
K, G	$\frac{9KG}{3K+G}$	$\frac{3K-2G}{2(3K+G)}$	G	$\frac{3K-2G}{3}$	K

$$D_{ijkl} = \frac{(1+\nu)}{2E}\left[-\frac{2\nu}{1+\nu}\delta_{ij}\delta_{kl} + \delta_{ik}\delta_{jl} + \delta_{il}\delta_{jk}\right] \tag{3.52}$$

$$C_{ijkl} = \frac{E}{2(1+\nu)}\left[\frac{2\nu}{1-2\nu}\delta_{ij}\delta_{kl} + \delta_{ik}\delta_{jl} + \delta_{il}\delta_{jk}\right] \tag{3.53}$$

3.11 Decomposition of Stress-Strain Relations

The linear isotropic elastic stress-strain relationships can be decomposed into two parts: a volumetric part relating mean stress σ_{kk} to volumetric strain ε_{kk}, and a deviatoric part relating deviatoric stress s_{ij} to deviatoric (distortion) strain e_{ij} in the following simple neat form

$$\sigma_{kk} = 3K\varepsilon_{kk}, \quad s_{ij} = 2G e_{ij} \tag{3.54}$$

$$\varepsilon_{kk} = \frac{1}{3K}\sigma_{kk}, \quad e_{ij} = \frac{1}{2G}s_{ij} \tag{3.55}$$

or

$$\sigma_{ij} = K\varepsilon_{kk}\delta_{ij} + 2G e_{ij} \tag{3.56}$$

$$\varepsilon_{ij} = \frac{1}{3K}p\,\delta_{ij} + \frac{1}{2G}s_{ij} \tag{3.57}$$

3.12 Basic Equations in Cylindrical and Spherical Coordinates

This section lists the basic equilibrium and compatibility equations for problems having cylindrical or spherical symmetry. The cylindrical and spherical coordinate systems are shown in Fig.3.3.

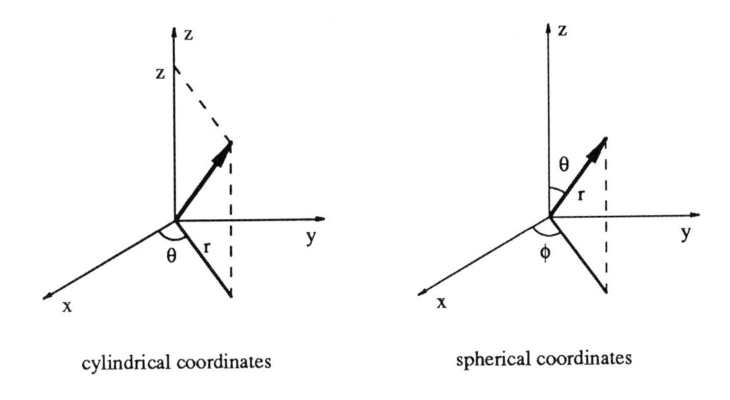

Figure 3.3. Cylindrical and spherical coordinate systems

3.12.1 Basic Equations for Problems Having Cylindrical Symmetric

For problems with cylindric symmetry, we have $\tau_{r\theta} = \tau_{\theta z} = 0$, $\varepsilon_{r\theta} = \varepsilon_{\theta z} = 0$, and $u_\theta = 0$. The basic equations are reduced to, in the usual engineering notation

$$\frac{\partial \sigma_r}{\partial r} + \frac{\partial \tau_{zr}}{\partial z} + \frac{\sigma_r - \sigma_\theta}{r} + F_r = 0$$

$$\frac{\partial \tau_{rz}}{\partial r} + \frac{\partial \sigma_z}{\partial z} + \frac{\tau_{rz}}{r} + F_z = 0 \tag{3.58}$$

$$\varepsilon_r = \frac{\partial u_r}{\partial r}, \quad \varepsilon_\theta = \frac{1}{r}\frac{\partial u_\theta}{\partial \theta} + \frac{u_r}{r}$$

$$\varepsilon_z = \frac{\partial u_z}{\partial z}, \quad \gamma_{rz} = \frac{\partial u_r}{\partial z} + \frac{\partial u_z}{\partial r} \tag{3.59}$$

3.12.2 Basic Equations for Problems Having Spherical Symmetric

For problems with spherical symmetry, we have $\sigma_\theta = \sigma_\phi$, $\tau_{r\theta} = \tau_{\theta\phi} = \tau_{\phi r} = 0$, $\varepsilon_\theta = \varepsilon_\phi$, $\varepsilon_{r\theta} = \varepsilon_{\theta\phi} = \varepsilon_{\phi r} = 0$, and $u_\theta = u_\phi = 0$. The stresses, strains, and displacements are single-variable functions of the coordinate r. The basic equations are further reduced to, in the usual engineering notation

$$\frac{d\sigma_r}{dr} + 2\frac{\sigma_r - \sigma_\theta}{r} + F_r = 0 \tag{3.60}$$

$$\varepsilon_r = \frac{du_r}{dr}, \quad \varepsilon_\theta = \varepsilon_\phi = \frac{u_r}{r} \tag{3.61}$$

Chapter 4

Yield Criteria

4.1 Representation of Yield Criteria

In uniaxial stress states, the elastic limit of a material is obtained by a well-defined yield stress point σ_0 on an actual stress-strain curve. In combined stress states, the elastic limit is defined mathematically by a certain *yield criterion* or *yield condition.* The initial yield criterion is a function of stress state σ_{ij} and can be generally expressed as

$$f(\sigma_{ij}) = 0 \tag{4.1}$$

The initial yield criterion generally contains several material constants to be fitted with available experimental results.

For isotropic materials, the stress state at a point can be uniquely represented by three principal stresses. Thus, the yield criterion for isotropic materials can be expressed as

$$f(\sigma_1, \sigma_2, \sigma_3) = 0 \tag{4.2}$$

Moreover, since principal stresses can be expressed in terms of either stress invariants, I_1, J_2, J_3, or the Haigh-Westergaard coordinates, ξ, ρ, θ, Eq.(4.2) can also be expressed as

$$f(I_1, J_2, J_3) = 0 \tag{4.3}$$

and

$$f(\xi, \rho, \theta) = 0 \tag{4.4}$$

Equations (4.2) to (4.4) represent a surface in the principal stress space. Such a surface is referred to as the *yield surface.* Within the yield surface, the material behaves elastically. On the yield surface, the material begins to yield.

The elastic-plastic behavior of most metallic materials is essentially hydrostatic pressure insensitive. This implies that yield criteria for this type of materials do not depend on I_1. For a hydrostatic pressure insensitive material, the yield criterion can generally be expressed as

$$f(J_2, J_3) = 0 \tag{4.5}$$

or

$$f(\rho, \theta) = 0 \tag{4.6}$$

The shape of a yield surface is best described by its cross-sectional shapes on deviatoric planes, (ρ, θ), and its meridians on meridian planes, (ξ, ρ). The cross-sections of a yield surface are the intersection curves between the yield surface and a deviatoric plane which is perpendicular to the hydrostatic axis ξ and with $\xi = \text{const}$. The meridians of a yield surface are the intersection curves between the surface and a meridian plane which contains the hydrostatic axis and with $\theta = \text{const}$.

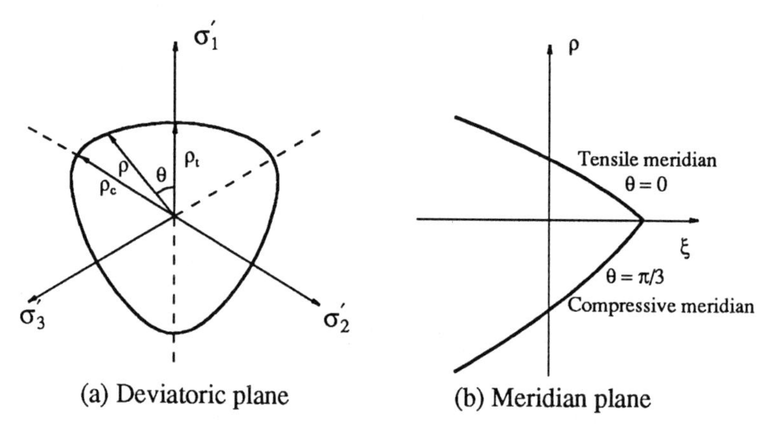

Figure 4.1. General shape of hydrostatic-pressure-dependent yield surface

Figure 4.1 shows the general shape of hydrostatic-pressure-dependent yield surfaces for isotropic materials. The cross-sectional shapes of this yield surface on deviatoric planes have threefold symmetry, Fig. 4.1a.

The meridian plane with $\theta = 0$, called *tensile meridian*, passes through the uniaxial tensile yield point. The meridian plane with $\theta = \pi/3$, called the *compressive meridian*, passes through the uniaxial compression yield point. The radius of a yield surface on the tensile meridian is denoted as ρ_t and on the compressive meridian is denoted as ρ_c. In Figure 4.1b, both the tensile and the compressive meridians of a general yield surface are shown.

For hydrostatic-pressure-independent yield surfaces, their meridians are straight lines parallel to the hydrostatic axis. This implies that shearing stress must be the major cause of yielding of this type of materials. Since the magnitude of shearing stress is important, not its direction, in governing yielding, it follows that the elastic-plastic behavior in tension and in compression should be equivalent for hydrostatic-pressure-independent materials. Thus, the cross-sectional shapes for this type of yield surfaces have six-fold symmetry, and $\rho_t = \rho_c$. This will be discussed further in the following section.

In many practical applications, some intersections of a yield surface in general stress space with particular planes are often of great interest. For example, the intersections of a yield surface with the (σ_1, σ_2) plane where $\sigma_3 = \text{const.}$, and with the (σ_x, τ_{xy}) plane (often referred to as (σ, τ) plane) where $\sigma_z = \sigma_3 = \text{const.}$ and $\sigma_y = \tau_{yz} = \tau_{xz} = 0$. These particular planes are called *stress sub-spaces.*

For brittle or granular materials, yield criterion is often referred to as failure criterion because for such materials yielding means fail. For example, the term failure criterion is often utilized for materials such as soils, concretes and rocks.

The material constants in a yield criterion must be determined by curve-fitting with simple tests, such as the uniaxial tension test and uniaxial compression test.

4.2 Hydrostatic-Pressure-Independent Materials

4.2.1 Tresca Criterion

The Tresca criterion states that yielding of a material would occur when the maximum shearing stress at a point of the material reaches a critical value k. In terms of principal stresses, we have

$$\text{Max}\left(\frac{1}{2}\,|\sigma_1 - \sigma_2|,\ \frac{1}{2}\,|\sigma_2 - \sigma_3|,\ \frac{1}{2}\,|\sigma_3 - \sigma_1|\right) = k \tag{4.7}$$

From a uniaxial tension test, we determine $k = \sigma_0/2$, and from a pure shear test, $k = \tau_0$. Thus, if the Tresca criterion is used, the tensile strength and the shear strength of a material are related by $\sigma_0 = 2\tau_0$. Figure 4.2 shows the Tresca yield surface on the π plane (the deviatoric plane with $\xi = 0$), on a meridian plane, on the (σ_1, σ_2) plane with $\sigma_3 = 0$, and on the (σ, τ) plane with $\sigma_3 = 0$.

The Tresca criterion can also be generally expressed as

$$2\sqrt{J_2}\,\sin\left(\theta + \frac{\pi}{3}\right) - \sigma_0 = 0\,, \quad 0 \le \theta \le \frac{\pi}{3} \tag{4.8}$$

where $\sigma_0 = 2k$. On the π plane, the yield criterion is expressed as

$$\rho = \sqrt{2J_2} = \frac{\sigma_0}{\sqrt{2}\,\sin\left(\theta + \frac{\pi}{3}\right)}\,, \quad 0 \le \theta \le \frac{\pi}{3} \tag{4.9}$$

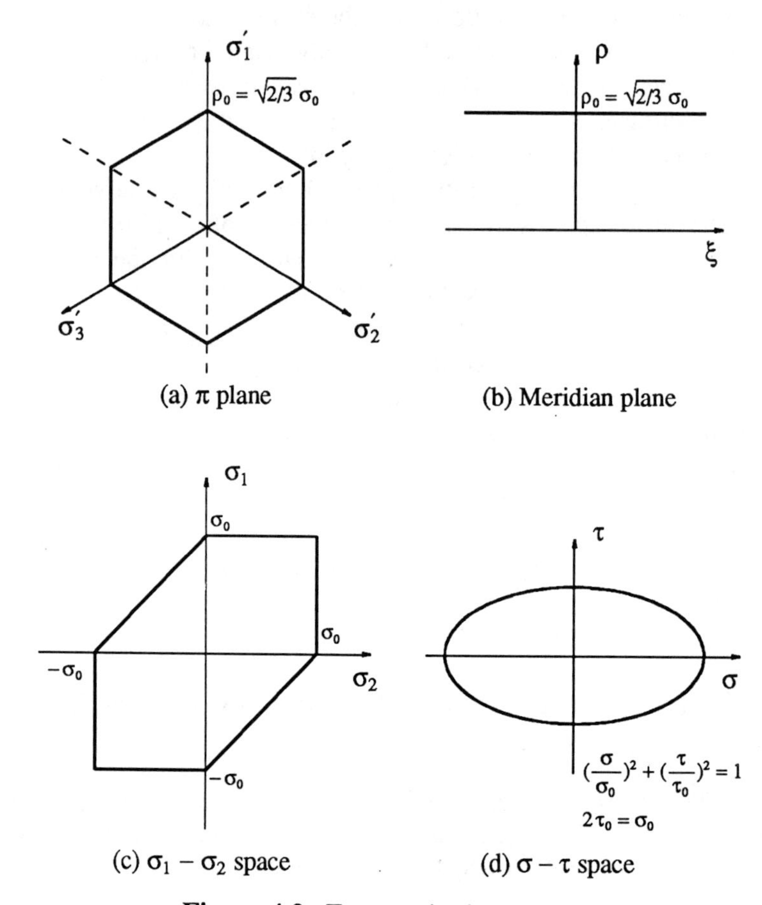

Figure 4.2. Tresca criterion

On a deviatoric plane, the Tresca criterion is a regular hexagon with six singular corners, and on a meridian plane, it is a straight line parallel to the ξ axis.

In the (σ_1, σ_2) sub-space with $\sigma_3 = 0$, the cross-section is an irregular hexagon, Fig. 4.2c. In the quarter $\sigma_1 \geq 0$, $\sigma_2 \leq 0$, the yield criterion is

$$\sigma_1 - \sigma_2 = \sigma_0 \tag{4.10}$$

In the $(\sigma - \tau)$ sub-space with $\sigma_3 = 0$, the yield surface is an ellipse

$$\sigma^2 + 4\tau^2 = \sigma_0^2 \tag{4.11}$$

or

$$\left[\frac{\sigma}{\sigma_0}\right]^2 + \left[\frac{\tau}{\tau_0}\right]^2 = 1$$

Since the Tresca criterion has a linear expression in the principal stress space, it is often employed for analytical solutions of elastic-plastic problems. However, the criterion does not take into account the effect of intermediate principal stress and contains singular corners causing possible troubles in numerical analysis.

4.2.2 von Mises Criterion

The von Mises criterion states that yielding of a material would occur when the maximum shearing strain energy at a point of the material reaches a critical value. Since the shear strain energy is proportional to the second invariant of the deviatoric stress tensor, J_2, the criterion can be expressed as

$$f(J_2) = J_2 - k^2 = 0 \tag{4.12}$$

From the uniaxial tension test, the constant k is determined as $k = \sigma_0 / \sqrt{3}$, and from a pure shear test, $k = \tau_0$. Thus, if the von Mises criterion is used, the tensile strength and the shear strength of a material are related by $\sigma_0 = \sqrt{3}\,\tau_0$. Thus, Eq.(4.12) can also be written as

$$J_2 - \frac{\sigma_0^2}{3} = 0 \tag{4.13}$$

Figure 4.3 shows the von Mises yield surface on the π plane, on a meridian plane, on the (σ_1, σ_2) plane with $\sigma_3 = 0$, and on the (σ, τ) plane with $\sigma_3 = 0$.

The cross-section of the yield surface with a deviatoric plane is a circle, as shown in Fig. 4.3a. The meridian of the surface is a straight line parallel to the hydrostatic pressure axis, Fig. 4.3b. On the π plane, the criterion is expressed as

$$\rho = \sqrt{\frac{2}{3}}\,\sigma_0 \tag{4.14}$$

In the (σ_1, σ_2) sub-space with $\sigma_3 = 0$, the criterion is an ellipse, Fig. 4.3c

$$\sigma_1^2 + \sigma_2^2 - \sigma_1\sigma_2 = \sigma_0^2 \tag{4.15}$$

or

$$\left(\frac{x}{\sqrt{2}\,\sigma_0}\right)^2 + \left(\frac{y}{\sqrt{2}\,\tau_0}\right)^2 = 1 \tag{4.16}$$

where

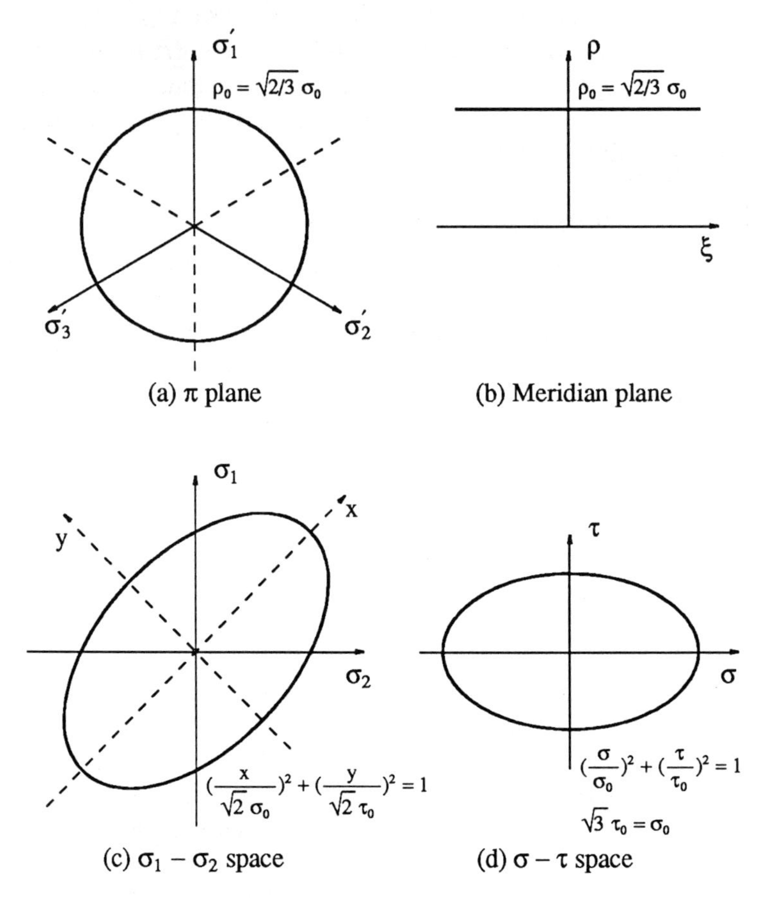

Figure 4.3. von Mises criterion

$$x = \frac{1}{\sqrt{2}}(\sigma_1 + \sigma_2), \quad y = \frac{1}{\sqrt{2}}(\sigma_2 - \sigma_1)$$

In the (σ, τ) sub-space with $\sigma_3 = 0$, the criterion is also an ellipse, Fig. 4.3c,

$$\left(\frac{\sigma}{\sigma_0}\right)^2 + \left(\frac{\tau}{\tau_0}\right)^2 = 1 \tag{4.17}$$

Since the von Mises criterion is of nonlinear form in terms of stress components, this criterion is somewhat harder to use for solving elastic-plastic problems.

4.3 Hydrostatic-Pressure-Dependent Materials

4.3.1 Rankine Criterion

The Rankine criterion states that yielding of a material would occur when the maximum principal stress at a point reaches the tensile strength of the material. In terms of principal stresses, it has the form

$$\sigma_1 = \sigma_0, \quad \sigma_2 = \sigma_0, \quad \sigma_3 = \sigma_0 \tag{4.18}$$

Figure 4.4 shows the Rankine yield surface on the π plane, on meridian planes, on the (σ_1, σ_2) plane with $\sigma_3 = 0$, and on the (σ, τ) plane with $\sigma_3 = 0$.

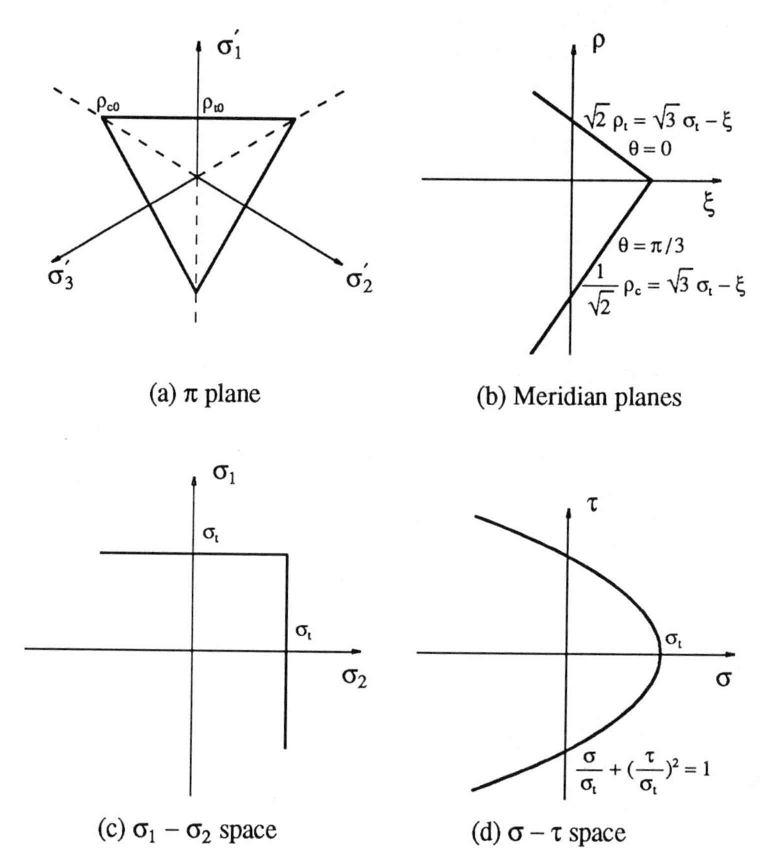

Figure 4.4. Rankine criterion

The general expression of the Rankine criterion is

$$2\sqrt{3J_2}\cos\theta + I_1 - 3\sigma_0 = 0\,, \qquad 0 \le \theta \le \frac{\pi}{3} \tag{4.19}$$

or

$$\sqrt{2}\rho\cos\theta + \xi - \sqrt{3}\,\sigma_0 = 0\,, \qquad 0 \le \theta \le \frac{\pi}{3} \tag{4.20}$$

The cross-section of the surface on the π plane is a regular triangle, Fig. 4.4a, given by

$$\rho_t = \frac{1}{\sqrt{2}}(\sqrt{3}\,\sigma_0 - \xi), \quad \rho_c = \sqrt{2}\,(\sqrt{3}\,\sigma_0 - \xi) \tag{4.21}$$

and $\rho_c = 2\rho_t$. The meridians of the surface are straight lines which intersect the ξ axis at the point $\xi_0 = \sqrt{3}\,\sigma_0$, Fig. 4.4b.

On the (σ_1, σ_2) plane, the surface is reduced to two straight lines, Fig. 4.4c,

$$\sigma_1 = \sigma_0\,, \quad \sigma_2 = \sigma_0 \tag{4.22}$$

On the (σ, τ) plane, the surface is reduced to a parabolic, Fig. 4.4d,

$$\frac{\sigma}{\sigma_0} + \left[\frac{\tau}{\sigma_0}\right]^2 = 1 \tag{4.23}$$

4.3.2 Mohr-Coulomb Criterion

The Mohr-Coulomb criterion can be considered as a generalization of the Tresca criterion. Both criteria assume that the maximum shearing stress determines the yielding of a material. However, the Tresca criterion assumes that the critical value of shearing stress is a constant, while the Mohr-Coulomb criterion considers the critical value of shearing stress on a plane to be a function of the normal stress acting on the same plane

$$|\,\tau\,| = c - \sigma\tan\phi \tag{4.24}$$

where c is the cohesion and ϕ the angle of internal friction; σ is the normal stress, and τ the resulting shearing stress on this plane. Both c and ϕ are material constants to be determined by experiments. Figure 4.5 shows the Mohr-Coulomb yield surface on the π plane, on meridian planes, on the (σ_1, σ_2) plane with $\sigma_3 = 0$, and on the (σ, τ) plane with $\sigma_3 = 0$.

The two parameters c and ϕ can be calibrated from two simple tests, e.g., a uniaxial tension test and a uniaxial compression test. Let σ_t be the tensile yield stress in the uniaxial tension, and σ_c the compression yield stress in the uniaxial compression test, we have

$$\sigma_t = \frac{2c\cos\phi}{1+\sin\phi}, \quad \sigma_c = \frac{2c\cos\phi}{1-\sin\phi} \tag{4.25}$$

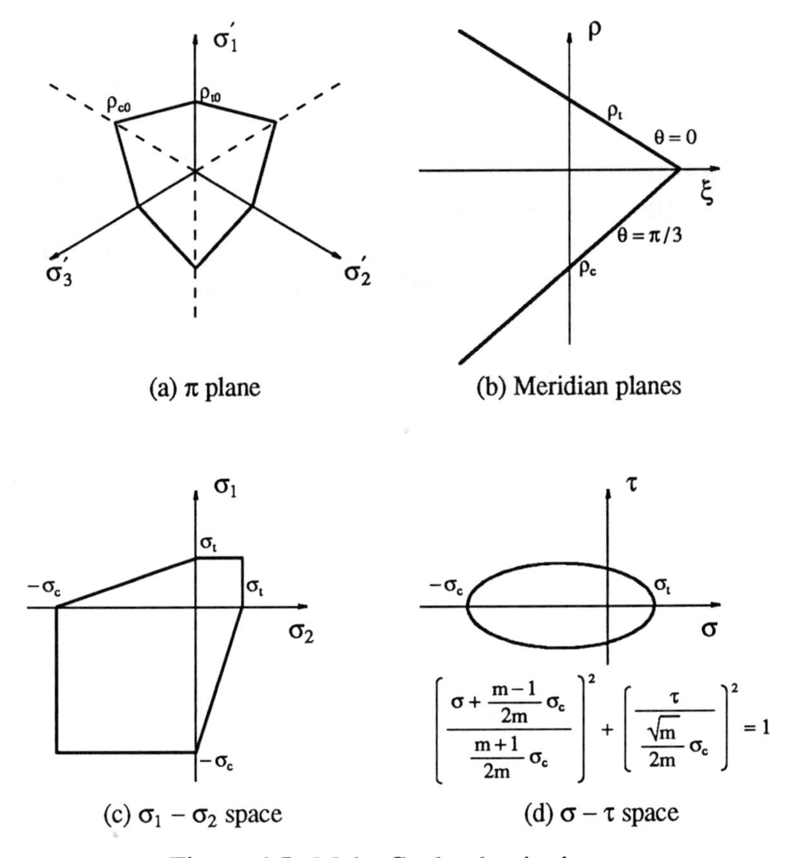

Figure 4.5. Mohr-Coulomb criterion

The ratio of σ_c and σ_t is defined as

$$m = \frac{\sigma_c}{\sigma_t} = \frac{1 + \sin\phi}{1 - \sin\phi} \tag{4.26}$$

In terms of m and σ_c, we have

$$\sin\phi = \frac{m-1}{m+1}, \quad \cos\phi = \frac{2\sqrt{m}}{m+1}, \quad 0 \le \phi \le \frac{\pi}{2} \tag{4.27}$$

and

$$c = \frac{\sigma_c}{2\sqrt{m}} \tag{4.28}$$

The general expression of the criterion has the form

$$\frac{1}{3}I_1 \sin\phi + \sqrt{J_2}\,\sin(\theta + \frac{\pi}{3}) + \sqrt{\frac{J_2}{3}}\,\cos(\theta + \frac{\pi}{3})\sin\phi - c\cos\phi = 0\,, \quad 0 \le \theta \le \frac{\pi}{3} \tag{4.29}$$

On the π plane, the cross-section of the surface is an irregular hexagon, Fig. 4.5a. The meridians of the surface are straight lines which intersect with the ξ axis at the point $\xi_0 = 2\sqrt{3}\,c/\tan\phi$, Fig. 4.5b. The two characteristic lengths of the surface on the deviatoric and meridian planes are

$$\rho_t = \frac{2\sqrt{6}\,c\cos\phi - 2\sqrt{2}\,\xi\sin\phi}{3 + \sin\phi} \tag{4.30}$$

$$\rho_c = \frac{2\sqrt{6}\,c\cos\phi - 2\sqrt{2}\,\xi\sin\phi}{3 - \sin\phi} \tag{4.31}$$

from which we have

$$\frac{\rho_t}{\rho_c} = \frac{3 - \sin\phi}{3 + \sin\phi} \tag{4.32}$$

On the (σ_1, σ_2) plane with $\sigma_3 = 0$, the surface is an irregular hexagon, Fig. 4.5c. In the quarter $\sigma_1 \ge 0$, $\sigma_2 \le 0$ of the plane, the criterion is expressed as

$$m\,\sigma_1 - \sigma_2 = \sigma_c \tag{4.33}$$

In the $\sigma - \tau$ sub-space, the surface is an ellipse, Fig. 4.5d

$$\left[\frac{\sigma + \frac{m-1}{2m}\sigma_c}{\frac{m+1}{2m}\sigma_c}\right]^2 + \left[\frac{\tau}{\frac{\sqrt{m}}{2m}\sigma_c}\right]^2 = 1 \tag{4.34}$$

4.3.3 Drucker-Prager criterion

The Drucker-Prager criterion is a simple extension of the von Mises criterion to include the effect of hydrostatic pressure on the yielding of materials. The extension is made by introducing an additional term that is proportional to I_1

$$f(J_2, J_3) = \alpha I_1 + \sqrt{J_2} - k = 0 \tag{4.35}$$

where α and k are material constants. From the uniaxial tension and uniaxial compression tests, we obtain

$$\sigma_t = \frac{\sqrt{3}\,k}{1+\sqrt{3}\,\alpha}, \quad \sigma_c = \frac{\sqrt{3}\,k}{1-\sqrt{3}\,\alpha} \tag{4.36}$$

Use the ratio $m = \sigma_c / \sigma_t$, we can also express the parameters α and k as

$$\alpha = \frac{m-1}{\sqrt{3}\,(m+1)}, \quad k = \frac{2\,\sigma_c}{\sqrt{3}\,(m+1)} \tag{4.37}$$

Figure 4.6 shows the yield surface on the π plane, on meridian planes, on the (σ_1, σ_2) plane with $\sigma_3 = 0$, and on the (σ, τ) plane with $\sigma_3 = 0$.

The Drucker-Prager can also be expressed generally as

$$\rho = \sqrt{2}\,k - \sqrt{6}\,\alpha\xi \tag{4.38}$$

On the deviatoric planes, the criterion is a circle with radius ρ, Fig. 4.6a. The meridians of the surface are straight lines which intersect with the ξ axis at the point $\xi_0 = k/\sqrt{3}\,\alpha$, Fig. 4.6b.

In the (σ_1, σ_2) sub-space with $\sigma_3 = 0$, the criterion is an ellipse, Fig. 4.6c,

$$\left[\frac{x + \dfrac{6\sqrt{2}\,k\alpha}{1-12\alpha^2}}{\dfrac{\sqrt{6}\,k}{1-12\alpha^2}}\right]^2 + \left[\frac{\dfrac{y}{\sqrt{2}\,k}}{\sqrt{1-12\alpha^2}}\right]^2 = 1 \tag{4.39}$$

where

$$x = \frac{1}{\sqrt{2}}(\sigma_1 + \sigma_2), \quad y = \frac{1}{\sqrt{2}}(\sigma_2 - \sigma_1)$$

In the (σ, τ) sub-space with $\sigma_3 = 0$, the criterion is also an ellipse, Fig. 4.6d,

$$\left[\frac{\sigma + \dfrac{3k\alpha}{1-3\alpha^2}}{\dfrac{\sqrt{3}\,k}{1-3\alpha^2}}\right]^2 + \left[\frac{\dfrac{\tau}{k}}{\sqrt{1-3\alpha^2}}\right]^2 = 1 \tag{4.40}$$

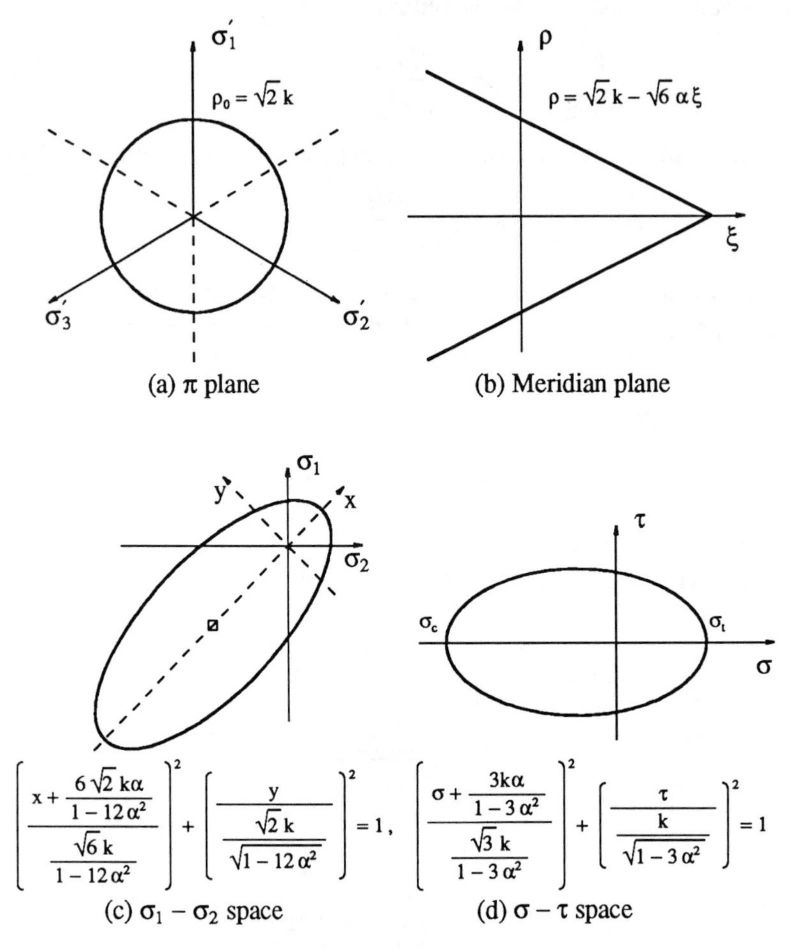

Figure 4.6. Drucker-Prager criterion

4.4 Tresca and von Mises Problems

Prob. 4.1 Show that the cross sectional shapes of a hydrostatic-pressure-independent isotropic yield surface in deviatoric planes have six-fold symmetry.

Prob. 4.2 Show that in the (σ_1, σ_2) plane:

a. von Mises ellipse circumscribes Tresca hexagon if the two surfaces are matched in the uniaxial tension test;

b. von Mises ellipse inscribes Tresca hexagon if the two surfaces are matched in the pure shear test.

Answer: Figure S4.2.

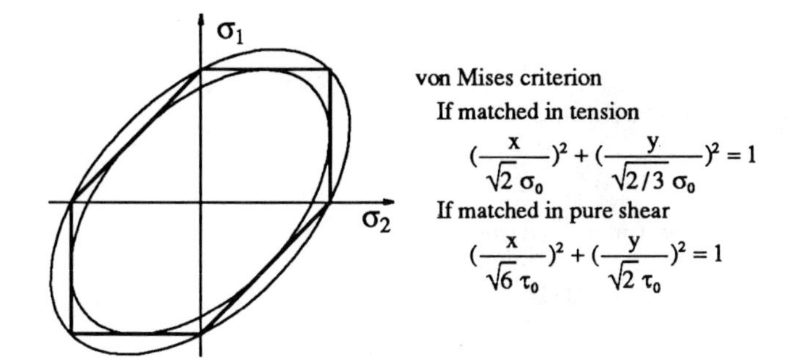

Figure S4.2.

Prob. 4.3: Draw the Tresca yield locus in the (σ_1, σ_2) plane in the case $\sigma_3 \neq 0$, and write the expression for each edge of the locus.

Answer: Figure S4.3.

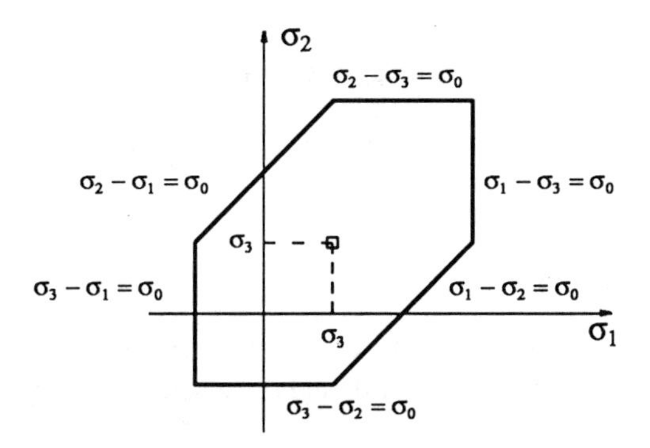

Figure S4.3.

Prob. 4.4 Derive the expression for Tresca criterion in the (σ, τ) subspace assuming $0 < \sigma_3 < \sigma_0$, and plot the corresponding yield locus on the (σ, τ) plane.

Solution: In the (σ, τ) sub-space, the principal stresses are

$$\sigma_1 = \frac{1}{2}(\sigma + \sqrt{\sigma^2 + 4\tau^2}) \geq 0, \quad \sigma_2 = \frac{1}{2}(\sigma - \sqrt{\sigma^2 + 4\tau^2}) \leq 0$$

If $\sigma_3 > \sigma_1$, or

$$\sigma_3\sigma + \tau^2 < \sigma_3^2$$

the Tresca criterion has the form

$$\sigma_3 - \sigma_2 = \sigma_0$$

Substitute the expression for σ_2, we obtain

$$\tau^2 - (\sigma_0 - \sigma_3)\sigma = (\sigma_0 - \sigma_3)^2$$

If $\sigma_3 < \sigma_1$, or

$$\sigma_3\sigma + \tau^2 > \sigma_3^2$$

the Tresca criterion has the form

$$\sigma_1 - \sigma_2 = \sigma_0$$

or

$$\sigma^2 + 4\tau^2 = \sigma_0^2$$

Thus, the locus of the Tresca criterion in the (σ, τ) plane for the case $0 < \sigma_3 < \sigma_0$ reduces to

$$\tau^2 - (\sigma_0 - \sigma_3)\sigma = (\sigma_0 - \sigma_3)^2, \quad \text{if } \sigma_3\sigma + \tau^2 < \sigma_3^2$$

$$\sigma^2 + 4\tau^2 = \sigma_0^2, \quad \text{if } \sigma_3\sigma + \tau^2 > \sigma_3^2$$

The two curves intersect at

$$\sigma = 2\sigma_3 - \sigma_0, \quad \tau = \pm\sqrt{\sigma_3(\sigma_0 - \sigma_3)}$$

Figure S4.4a shows the yield locus drawn in thicker lines. Figure S4.4b shows the yield loci for the cases $r = 1.5$, 1.0, and 0.5 where $r = \sigma_3 / \tau_0$.

Prob. 4.5 Same as Prob. 4.4 but with the condition $-\sigma_0 < \sigma_3 < 0$.

Answer:

$$\tau^2 + (\sigma_0 + \sigma_3)\sigma = (\sigma_0 + \sigma_3)^2, \quad \text{if } \sigma_3\sigma + \tau^2 < \sigma_3^2$$

$$\sigma^2 + 4\tau^2 = \sigma_0^2, \quad \text{if } \sigma_3\sigma + \tau^2 > \sigma_3^2$$

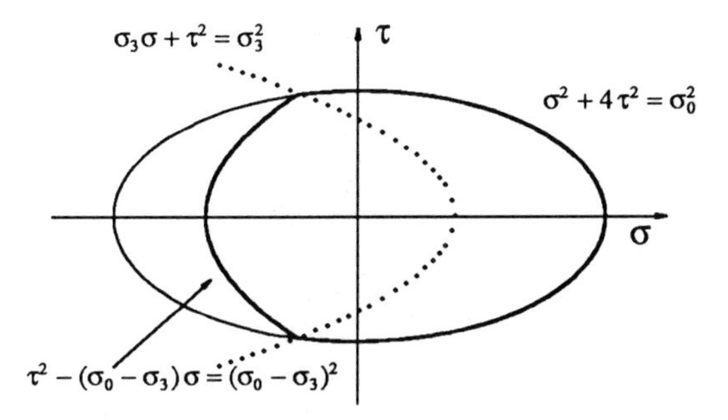

Figure S4.4(a).

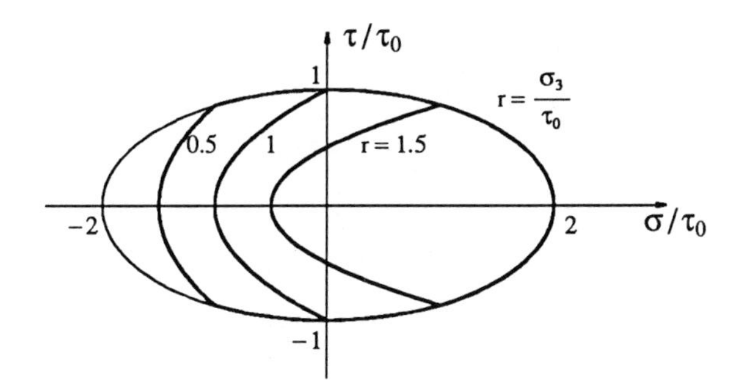

Figure S4.4(b).

They intersect at

$$\sigma = 2\sigma_3 + \sigma_0, \quad \tau = \pm\sqrt{-\sigma_3(\sigma_0 + \sigma_3)}$$

Figure S4.5a shows the yield locus drawn in thicker lines. Figure S4.5b shows the yield loci for the cases r = 1.5, 1.0, and 0.5 where $r = -\sigma_3/\tau_0$.

Prob. 4.6 Show that in the (σ_1, σ_2) sub-space, if $\sigma_3 = \text{const.} \neq 0$, von Mises criterion can be expressed as

$$\left[\frac{x - \sqrt{2}\,\sigma_3}{\sqrt{6}\,k}\right]^2 + \left[\frac{y}{\sqrt{2}\,k}\right]^2 = 1$$

where

$$x = \frac{1}{\sqrt{2}}(\sigma_2 + \sigma_1), \quad y = \frac{1}{\sqrt{2}}(\sigma_2 - \sigma_1)$$

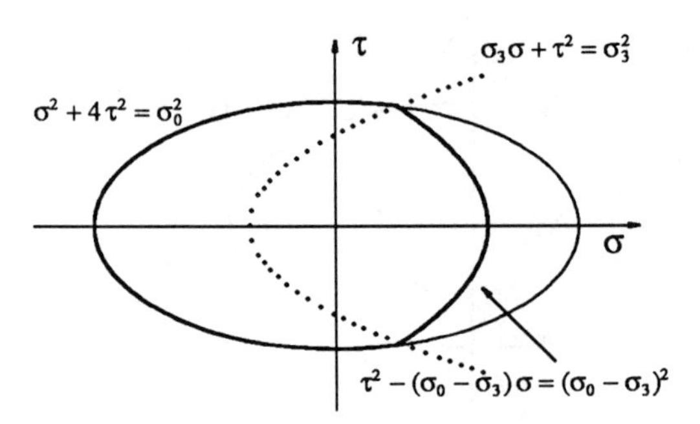

Figure S4.5(a).

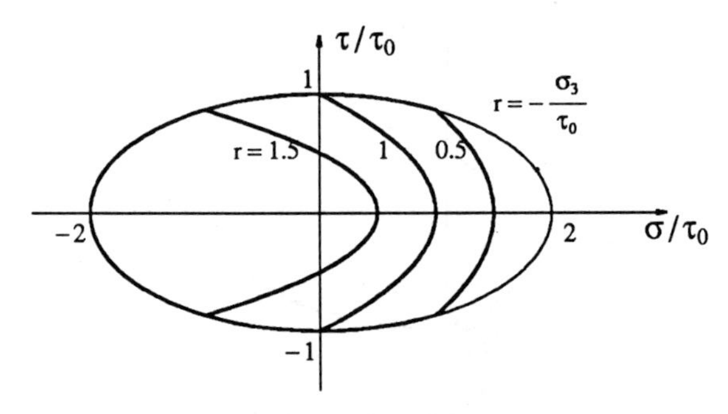

Figure S4.5(b).

Plot the surface in the (σ_1, σ_2) plane for the cases $\sigma_3 > 0$, $\sigma_3 = 0$ and $\sigma_3 < 0$. What is the range of σ_3 for which the expression remains valid.

Answer: Figure S4.6.

Prob. 4.7 Show that in the (σ, τ) sub-space, if $\sigma_3 = \text{const.} \neq 0$, von Mises criterion has the form

$$\left[\frac{\sigma - \frac{1}{2}\sigma_3}{\sqrt{3}\sqrt{k^2 - \frac{1}{4}\sigma_3^2}}\right]^2 + \left[\frac{\tau}{\sqrt{k^2 - \frac{1}{4}\sigma_3^2}}\right]^2 = 1$$

Plot the surface in the (σ, τ) plane for the cases $\sigma_3 > 0$, $\sigma_3 = 0$ and $\sigma_3 < 0$. What is the range of σ_3 for which the expression remains valid.

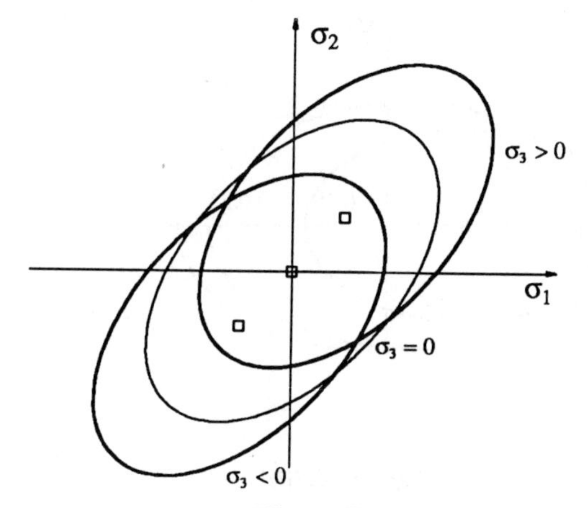

Figure S4.6.

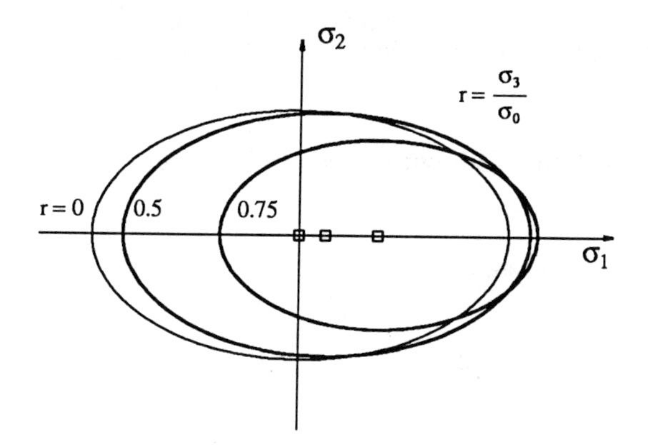

Figure S4.7(a).

Answer: Figures S4.7a and S4.7b.

Prob. 4.8 A material yields at the stress state

$$\sigma_1 = \sigma_0, \quad \sigma_2 = \sigma_0, \quad \sigma_3 = 0$$

Using (a) von Mises criterion and (b) Tresca criterion, find the strengths of the material in uniaxial tension and pure shear, and the yield value of p for the following loading path: $\sigma_1 = 2\,\text{p}$, $\sigma_2 = \text{p}$, $\sigma_3 = 3\,\text{p}$.

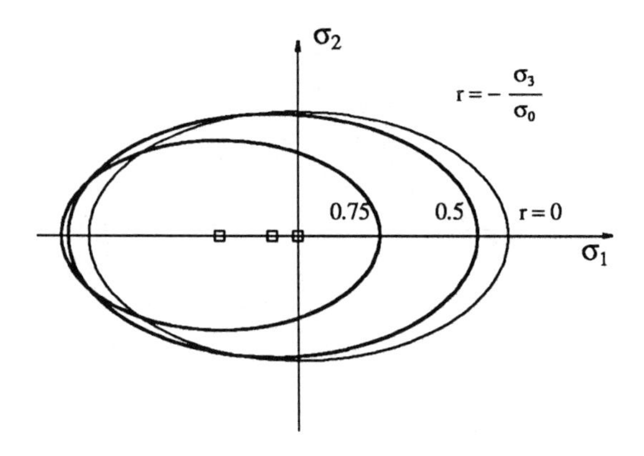

Figure S4.7(b).

Answer:

a. tensile strength $\sigma_t = \sigma_0$, shearing strength $\tau_0 = \sigma_0 / \sqrt{3}$, and $p_y = \sigma_0 / \sqrt{3}$.
b. tensile strength $\sigma_t = \sigma_0$, shearing strength $\tau_0 = \sigma_0 / 2$, and $p_y = \sigma_0 / 2$.

Prob. 4.9 Given the following two yield states of biaxial loading condition (i.e., $\sigma_3 = 0$),

$$(1) \quad \sigma_1 = \sigma, \quad \sigma_2 = \frac{\sigma}{3}$$

$$(2) \quad \sigma_1 = \sigma, \quad \sigma_2 = \frac{3}{2}\sigma$$

determine the tensile and shearing strengths corresponding to each yield state based on (a) von Mises criterion, and (b) Tresca criterion. Plot von Mises and Tresca yield surfaces corresponding to each yield state in the (σ_1, σ_2) plane, and mark these matched points on the surfaces.

Answer:

(1) Figure S4.9a shows the surfaces.

Tresca criterion: $\sigma_t = \sigma$, $\tau_0 = \sigma / 2$

von Mises criterion: $\sigma_t = \sqrt{\frac{7}{9}}\,\sigma$, $\tau_0 = \sqrt{\frac{7}{27}}\,\sigma$

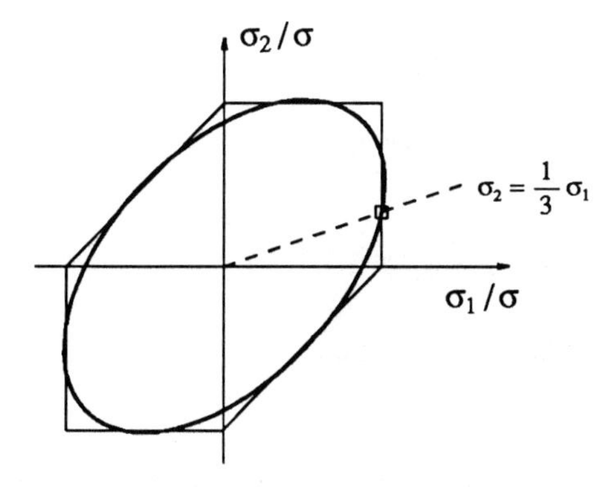

Figure S4.9(a).

(2) Figure S4.9b shows the surfaces

Tresca criterion: $\sigma_t = \dfrac{3}{2}\sigma,\ \tau_0 = \dfrac{3}{4}\sigma$

von Mises criterion: $\sigma_t = \sqrt{\dfrac{7}{4}}\ \sigma,\ \tau_0 = \sqrt{\dfrac{7}{12}}\ \sigma$

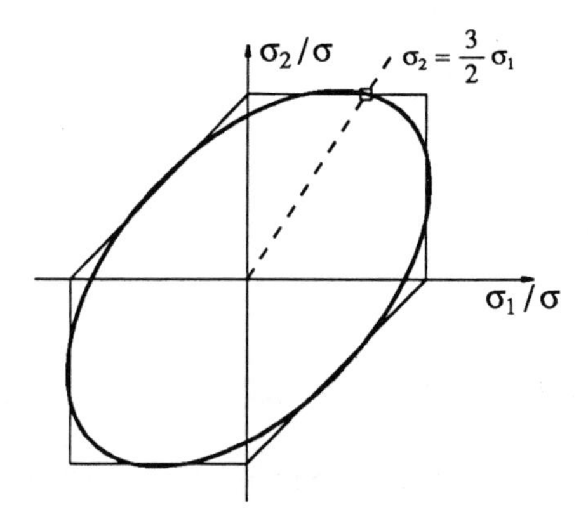

Figure S4.9(b).

Prob. 4.10 Show that the following expressions always satisfy von Mises criterion.

$$\sigma_1 = \frac{2}{3}\sigma_s \cos(\phi - \frac{\pi}{3}) + \sigma_0$$

$$\sigma_2 = \frac{2}{3}\sigma_s \cos(\phi + \frac{\pi}{3}) + \sigma_0$$

$$\sigma_3 = -\frac{2}{3}\sigma_s \cos\phi + \sigma_0$$

If $\sigma_1 \geq \sigma_2 \geq \sigma_3$, what is the range of ϕ.

Prob. 4.11 Show that von Mises criterion can be expressed as

$$I_1^2 - 3 I_2 = \sigma_0^2$$

where I_1, and I_2 are the first and second invariants of stress tensor.

Prob. 4.12 Derive the expressions for Tresca and von Mises criteria under the plane strain condition ($\varepsilon_z = 0$, $\gamma_{yz} = \gamma_{zx} = 0$) in terms of σ_x, σ_y, and τ_{xy}. Assume $\nu = \frac{1}{2}$, and match the two criteria in uniaxial tension.

Answer:

Tresca criterion: $(\sigma_x - \sigma_y)^2 + 4\,\tau_{xy} = \sigma_0^2$

von Mises criterion: $(\sigma_x - \sigma_y)^2 + 4\,\tau_{xy} = \frac{4}{3}\,\sigma_0^2$

Prob. 4.13 Derive the expressions for Tresca and von Mises criteria under the plane stress condition ($\sigma_z = 0$, $\tau_{yz} = \tau_{zx} = 0$) in terms of σ_x, σ_y, and τ_{xy}. Match the two criteria in uniaxial tension.

Answer:

Tresca criterion:

$(\sigma_x - \sigma_0)(\sigma_y - \sigma_0) = \tau_{xy}^2$ if $\sigma_x + \sigma_y > 0$ and $\sigma_x\sigma_y \geq \tau_{xy}^2$

$(\sigma_x - \sigma_y)^2 + 4\,\tau_{xy} = \sigma_0^2$ if $\sigma_x\sigma_y \leq \tau_{xy}^2$

$$(\sigma_x + \sigma_0)(\sigma_y + \sigma_0) = \tau_{xy}^2 \qquad \text{if } \sigma_x + \sigma_y < 0 \text{ and } \sigma_x\sigma_y \geq \tau_{xy}^2$$

von Mises criterion:

$$\sigma_x^2 + \sigma_y^2 - \sigma_x\sigma_y + 3\,\tau_{xy} = \sigma_0^2$$

4.5 Mohr-Coulomb Problems

Prob. 4.14 Show that

a. When the ratio of the two characteristic lengths, ρ_t/ρ_c, of Mohr-Coulomb criterion is 1, the criterion reduces to Tresca criterion;

b. When the ratio is $\frac{1}{2}$, the criterion reduces to Rankine criterion.

Prob. 4.15 Draw the Mohr-Coulomb yield locus in the (σ_1, σ_2) plane for the case $\sigma_3 = \text{const.} \neq 0$, and derive the expressions for each edge of the yield locus.

Answer: Figure S4.15.

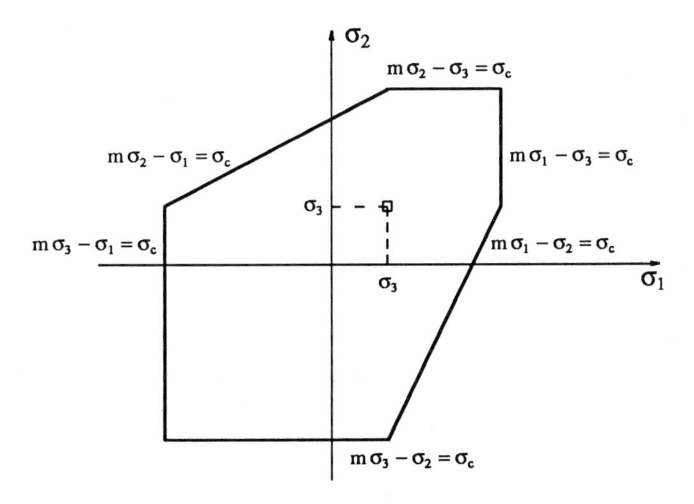

Figure S4.15.

Prob. 4.16 Derive the expression for Mohr-Coulomb criterion in the (σ, τ) sub-space for the case $\sigma_3 = 0$ (Eq.(4.34)), and plot the yield locus in the (σ, τ) plane for the cases m = 1, 2, 4, 8.

Answer: Figure S4.16.

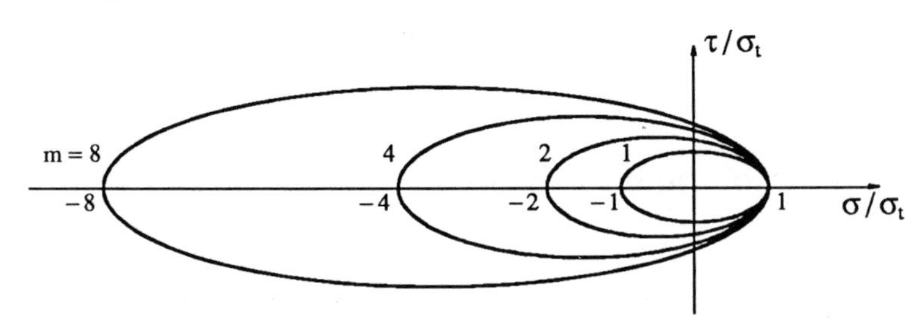

Figure S4.16.

Prob. 4.17 Show that in the (σ, τ) plane for the case $0 < \sigma_3 < \sigma_t$, the locus of Mohr-Coulomb criterion consists of the following two curves

$$\tau^2 - (\sigma_c - m\,\sigma_3)\,\sigma = (\sigma_c - m\,\sigma_3)^2, \qquad \text{if } \sigma_3\sigma + \tau^2 < \sigma_3^2$$

$$\left(\frac{\sigma + \dfrac{m-1}{2m}\sigma_c}{\dfrac{m+1}{2m}\sigma_c}\right)^2 + \left(\frac{\tau}{\dfrac{\sqrt{m}}{2m}\sigma_c}\right)^2 = 1, \quad \text{if } \sigma_3\sigma + \tau^2 > \sigma_3^2$$

and they intersect at

$$\sigma = (1+m)\,\sigma_3 - \sigma_c, \quad \tau = \pm\sqrt{\sigma_3(\sigma_c - m\,\sigma_3)}$$

Answer: Figure S4.17a shows the yield locus drawn in thicker lines. Figure S4.17b shows the yield loci for the cases $r = 0.3, 0.25, 0.15$ and 0.05 where $r = \sigma_3/\sigma_c$.

Prob. 4.18 Show that in the (σ, τ) plane for the case $-\sigma_c < \sigma_3 < 0$, the locus of Mohr-Coulomb criterion consists of the following two curves

$$m^2\tau^2 + m\,(\sigma_c + \sigma_3)\,\sigma = (\sigma_c + \sigma_3)^2, \qquad \text{if } \sigma_3\sigma + \tau^2 < \sigma_3^2$$

$$\left(\frac{\sigma + \dfrac{m-1}{2m}\sigma_c}{\dfrac{m+1}{2m}\sigma_c}\right)^2 + \left(\frac{\tau}{\dfrac{\sqrt{m}}{2m}\sigma_c}\right)^2 = 1, \quad \text{if } \sigma_3\sigma + \tau^2 > \sigma_3^2$$

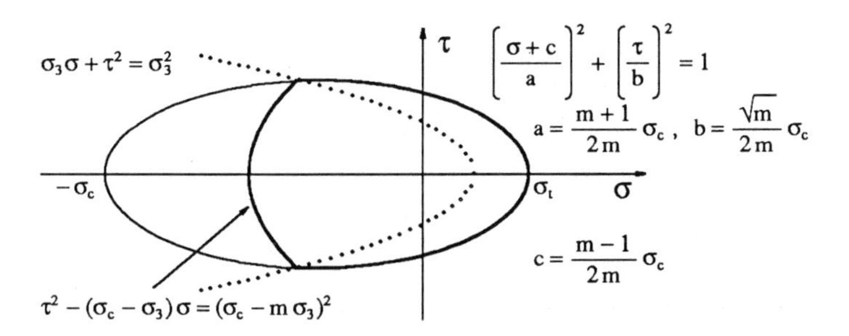

Figure S4.17(a).

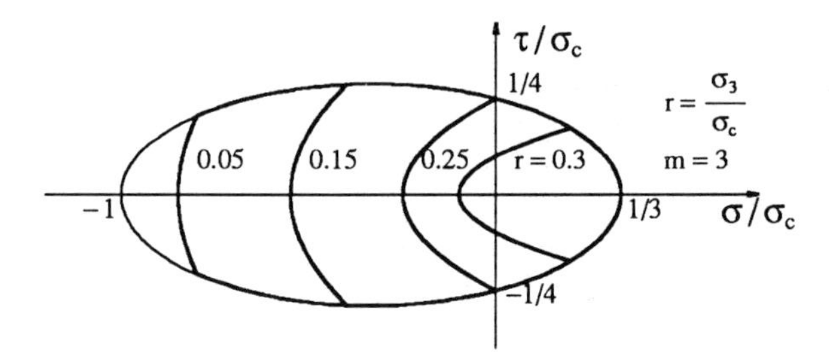

Figure S4.17(b).

and they intersect at

$$\sigma = \frac{(1+m)}{m}\sigma_3 + \frac{1}{m}\sigma_c, \quad \tau = \pm\sqrt{-\frac{\sigma_3}{m}(\sigma_c + \sigma_3)}$$

Answer: Figure S4.18a shows the yield locus drawn in thicker lines. Figure S4.18b shows the yield loci for the cases $r = 0.25, 0.50$ and 0.75 where $r = -\sigma_3/\sigma_c$.

Prob. 4.19 A material yields under the compression loading with $\sigma_1 = \sigma_2 = -\frac{1}{2}\sigma_c$, $\sigma_3 = -3\sigma_c$, where σ_c is the uniaxial compression strength of the material.

a. Determine the constants c and ϕ in terms of σ_c for Mohr-Coulomb criterion;

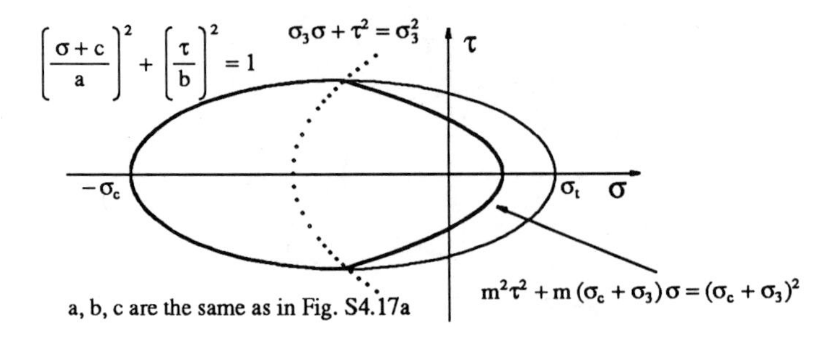

Figure S4.18(a).

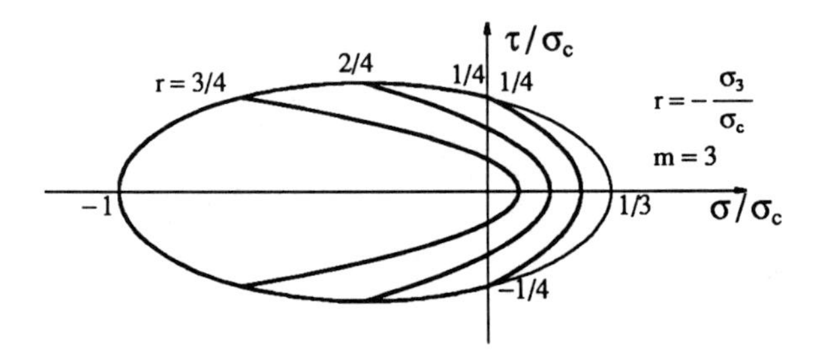

Figure S4.18(b).

b. Find the pure shear and uniaxial tensile strengths by Mohr-Coulomb criterion.
c. Predict the yield stress p_y for the loading path $\sigma_x = p$, $\sigma_y = 2p$, $\tau_{xy} = \frac{1}{2}p$, all other stress components are zero.

Answer:

(a) $c = \frac{1}{4}\sigma_c, \quad \tan\phi = \frac{3}{4}, \quad \text{or } \phi = 36.87^\circ$

(b) $\sigma_t = \frac{1}{4}\sigma_c, \quad \tau_0 = \frac{1}{5}\sigma_c$

(c) $p_y = \dfrac{\sigma_0}{2(3+\sqrt{2})}$

4.6 Drucker-Prager Problems

Prob. 4.20 Show that in the (σ_1, σ_2) sub-space, with $\sigma_3 = 0$, Drucker-Prager criterion is reduced to (Eq.(4.39))

$$\left[\frac{x + \frac{6\sqrt{2}\,k\alpha}{1-12\alpha^2}}{\frac{\sqrt{6}\,k}{1-12\alpha^2}}\right]^2 + \left[\frac{y}{\frac{\sqrt{2}\,k}{\sqrt{1-12\alpha^2}}}\right]^2 = 1$$

where

$$x = \frac{1}{\sqrt{2}}(\sigma_1 + \sigma_2)\,, \quad y = \frac{1}{\sqrt{2}}(\sigma_2 - \sigma_1)$$

Plot the yield locus in the (σ_1, σ_2) plane for the cases $\alpha = 0,\ 0.1,\ 0.2$.

Answer: Figure S4.20.

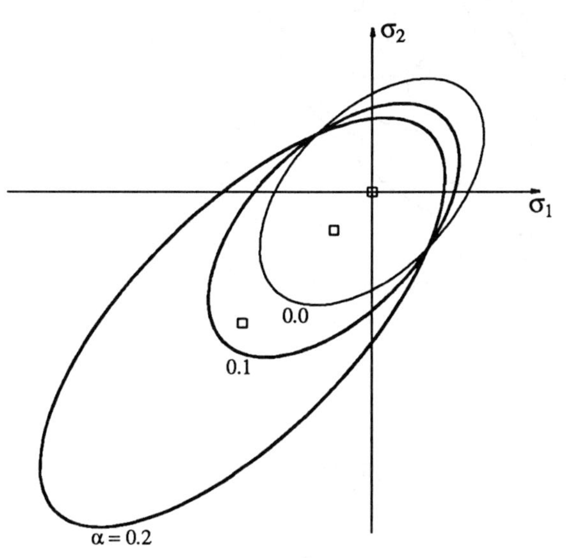

Figure S4.20.

Prob. 4.21 Show that if $\sigma_3 = \text{const.} \neq 0$ Drucker-Prager criterion in the (σ_1, σ_2) sub-space s reduced to

$$\left[\frac{x + \frac{\sqrt{2}\,[6\,\alpha k - (1+6\,\alpha^2)\,\sigma_3]}{1-12\alpha^2}}{\frac{\sqrt{6}\,(k - 3\,\alpha\,\sigma_3)}{1-12\alpha^2}}\right]^2 + \left[\frac{y}{\frac{\sqrt{2}\,(k - 3\,\alpha\,\sigma_3)}{\sqrt{1-12\alpha^2}}}\right]^2 = 1$$

where

$$x = \frac{1}{\sqrt{2}}(\sigma_1 + \sigma_2), \quad y = \frac{1}{\sqrt{2}}(\sigma_2 - \sigma_1)$$

Plot the yield locus in the (σ_1, σ_2) plane for the cases $\dfrac{\sigma_3}{k} = 0.5$, 0.0, −0.5 ,−1.0 and $\alpha = 0.1$.

Answer: Figure S4.21.

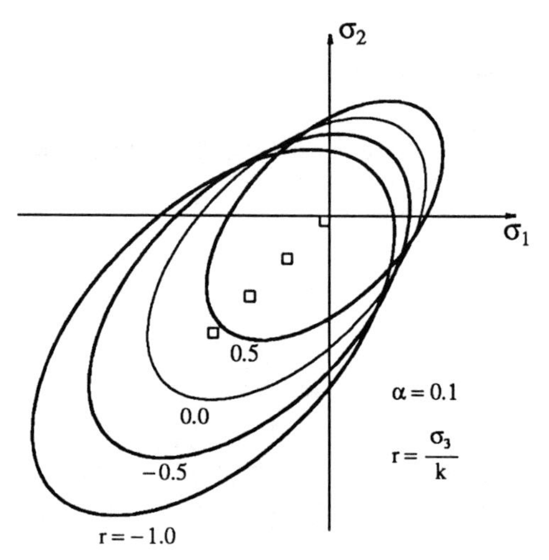

Figure S4.21.

Prob. 4.22 Show that in the (σ, τ) sub-space with $\sigma_3 = 0$, Drucker-Prager criterion is reduced to (Eq.(4.40))

$$\left[\frac{\sigma + \dfrac{3k\alpha}{1-3\alpha^2}}{\dfrac{\sqrt{3}k}{1-3\alpha^2}}\right]^2 + \left[\frac{\dfrac{\tau}{k}}{\sqrt{1-3\alpha^2}}\right]^2 = 1$$

Plot the yield locus in the (σ, τ) plane for the cases of $\alpha = 0$, 0.1, 0.2.

Answer: Figure S4.22.

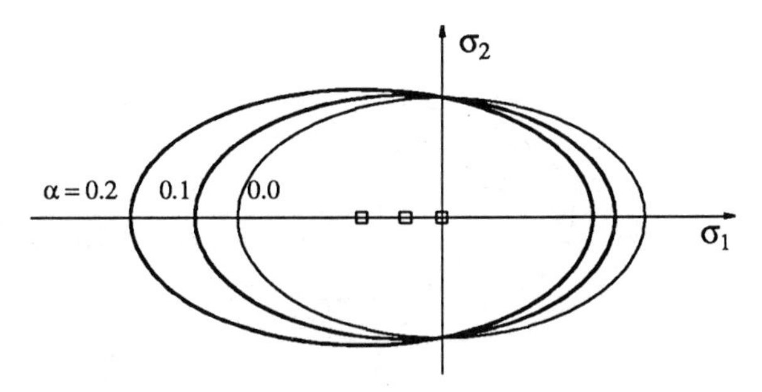

Figure S4.22.

Prob. 4.23 Show that in the (σ, τ) sub-space, if $\sigma_3 = \text{const.} \neq 0$, Drucker-Prager criterion is reduced to

$$\left[\frac{\sigma + \dfrac{3\alpha(k - \alpha\sigma_3) - \dfrac{1}{2}\sigma_3}{1 - 3\alpha^2}}{\dfrac{\sqrt{3}\sqrt{k^2 - \dfrac{1}{4}\sigma_3^2 - 3\alpha\sigma_3(k - \alpha\sigma_3)}}{1 - 3\alpha^2}}\right]^2 + \left[\frac{\tau}{\dfrac{\sqrt{k^2 - \dfrac{1}{4}\sigma_3^2 - 3\alpha\sigma_3(k - \alpha\sigma_3)}}{\sqrt{1 - 3\alpha^2}}}\right]^2 = 1$$

Plot the yield locus in the (σ, τ) plane for the cases of $\dfrac{\sigma_3}{k} = 0.8, 0.4, 0.0, -1.0, -2.0$ and $\alpha = 0.2$.

Answer: Figures S4.23a, and S4.23b.

Prob. 4.24 A material fails under the biaxial compression test with $\sigma_1 = \sigma_2 = -\sigma_{bc}$, $\sigma_3 = 0$, where $\sigma_{bc} = \dfrac{3}{2}\sigma_c$ is the biaxial compression strength, and σ_c is the uniaxial compression strength.

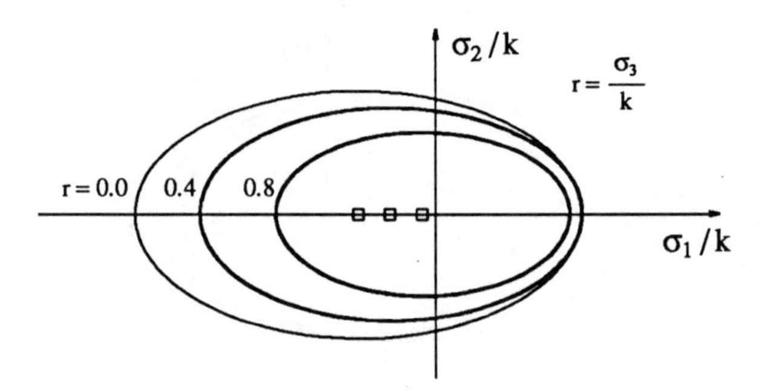

Figure S4.23(a).

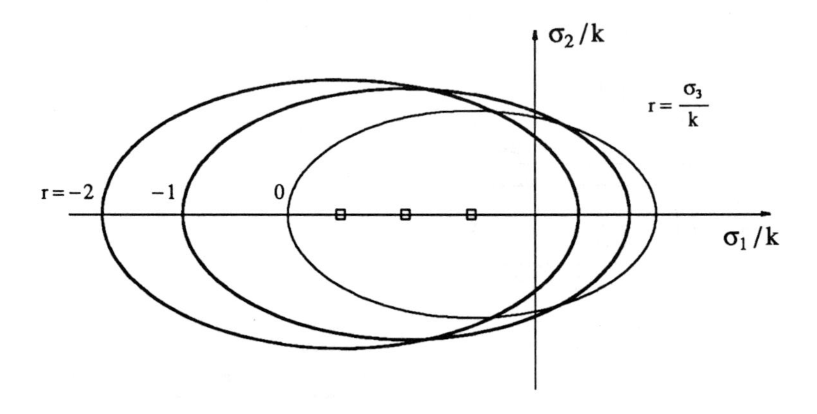

Figure S4.23(b).

a. Determine the constants α and k in terms of σ_c for Drucker-Prager criterion;
b. Find the pure shear and uniaxial tensile strengths by Drucker-Prager criterion.
c. Predict the yield stress p_y for the loading path $\sigma_x = p$, $\sigma_y = 2p$, $\tau_{xy} = \frac{1}{2}p$, all other stress components are zero.

Answer:

(a) $\alpha = \frac{\sqrt{3}}{12}, \quad k = \frac{\sqrt{3}}{4}\sigma_c$

(b) $\sigma_t = \frac{3}{5}\sigma_c, \quad \tau_0 = \frac{\sqrt{3}}{4}\sigma_c$

(c) $p_y = \dfrac{\sqrt{3}}{3+2\sqrt{5}}\sigma_c$

4.7 Tension Cut-Off Problems

Prob. 4.25 Assume that the material yields when the maximum principal strain value equals to the yield strain in uniaxial tension (or uniaxial compression), $\varepsilon_0 = \sigma_0 / E$. Express the yield criterion in terms of principal stresses, and plot the yield surface on the (σ_1, σ_2) plane with $\sigma_3 = 0$.

Answer: The yield surface in the (σ_1, σ_2) plane is plotted in Fig. S4.25.

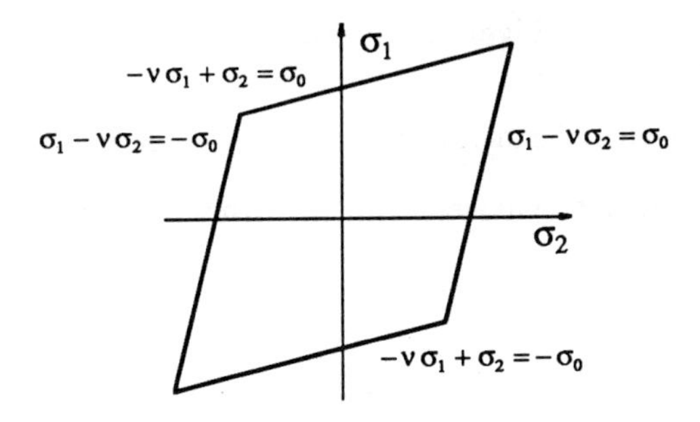

Figure S4.25.

Prob. 4.26 Derive the expression for Rankine criterion in the (σ, τ) sub-space [Eq. (4.23)].

Prob.4.27 The Mohr-Coulomb criterion with a tension cutoff by Rankine criterion is a combined criterion. The parameter m for Mohr-Coulomb criterion is 3. However, the actual ratio of σ_c / σ_t is 9. Plot the combined surface in:

a. The meridian planes $\theta = 0$ and $\theta = \pi/3$;
b. The (σ_1, σ_2) plane with $\sigma_3 = 0$;
c. The (σ, τ) plane with $\sigma_3 = 0$.

Answer:

$$\frac{\sigma_c}{c} = 2\sqrt{3}, \quad \frac{\sigma_t}{c} = \frac{2\sqrt{3}}{9}$$

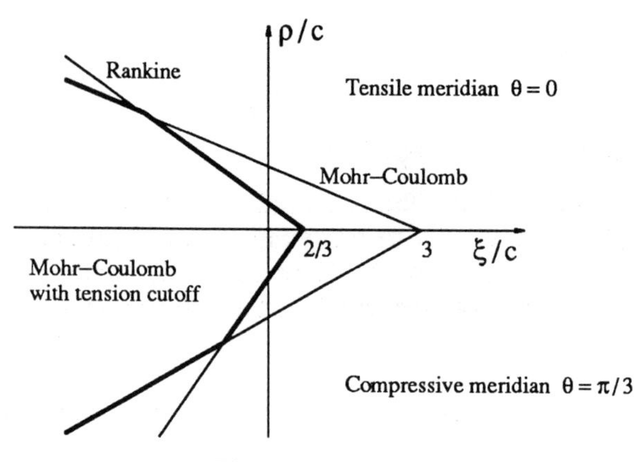

Figure S4.27(a).

(a) The combined surface in the meridian planes is shown in Fig. S4.27a.

The Mohr-Coulomb criterion in the meridian planes has the form

$$\frac{\rho_t}{c} = \frac{\sqrt{2}}{7}(6 - 2\frac{\xi}{c}), \quad \frac{\rho_c}{c} = \frac{\sqrt{2}}{5}(6 - 2\frac{\xi}{c}),$$

The Rankine criterion in the meridian planes has the form

$$\frac{\rho_t}{c} = \frac{1}{\sqrt{2}}(\frac{2}{3} - \frac{\xi}{c}), \quad \frac{\rho_c}{c} = \sqrt{2}\,(\frac{2}{3} - \frac{\xi}{c}),$$

They intersect at

tensile meridian: $\frac{\xi}{c} = -\frac{22}{9}, \quad \frac{\rho}{c} = \frac{14}{9}\sqrt{2}$

compressive meridian: $\frac{\xi}{c} = -\frac{8}{9}, \quad \frac{\rho}{c} = \frac{14}{9}\sqrt{2}$

(b) The combined surface in the (σ_1, σ_2) plane is shown in Fig. S4.27b.

(c) The combined surface in the (σ, τ) plane is shown in Fig. S4.27c.

The Mohr-Coulomb criterion in thc (σ, τ) plane is reduced to

$$\left[\frac{\sigma + \frac{1}{3}\sigma_c}{\frac{2}{3}\sigma_c}\right]^2 + \left[\frac{\tau}{\frac{\sqrt{3}}{6}\sigma_c}\right]^2 = 1$$

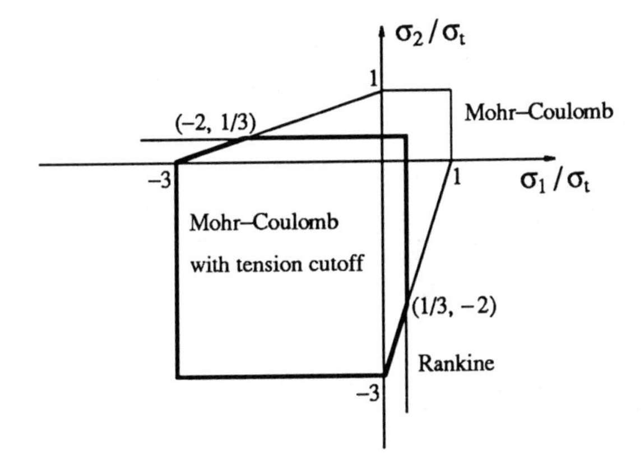

Figure S4.27(b).

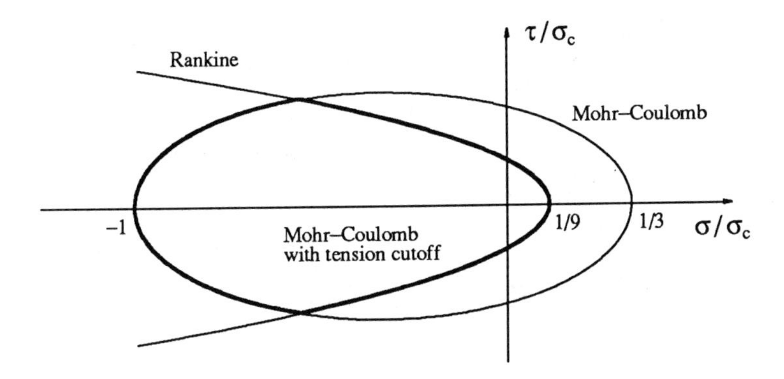

Figure S4.27(c).

The Rankine criterion in the (σ, τ) plane is reduced to

$$9\frac{\sigma}{\sigma_c} + 81\left(\frac{\tau}{\sigma_c}\right)^2 = 1$$

They intersect at

$$\frac{\sigma}{\sigma_c} = -\frac{5}{9}, \quad \frac{\tau}{\sigma_c} = \pm\frac{\sqrt{2}}{3\sqrt{3}}$$

Prob.4.28 The Drucker-Prager criterion with a tension cutoff by Rankine criterion is a combined criterion. The parameter α for Drucker-Prager criterion is $\sqrt{3}/8$. The actual ratio of σ_c/σ_t is 8. Plot the combined surface in:

a. The meridian planes $\theta = 0$ and $\theta = \pi/3$;
b. The (σ_1, σ_2) plane with $\sigma_3 = 0$;
c. The (σ, τ) plane with $\sigma_3 = 0$.

Answer:

$$\frac{\sigma_c}{k} = \frac{8\sqrt{3}}{5}, \quad \frac{\sigma_t}{k} = \frac{\sqrt{3}}{5}, \quad m = \frac{11}{5}$$

(a) The combined surface in the meridian planes is shown in Fig. S4.28a.

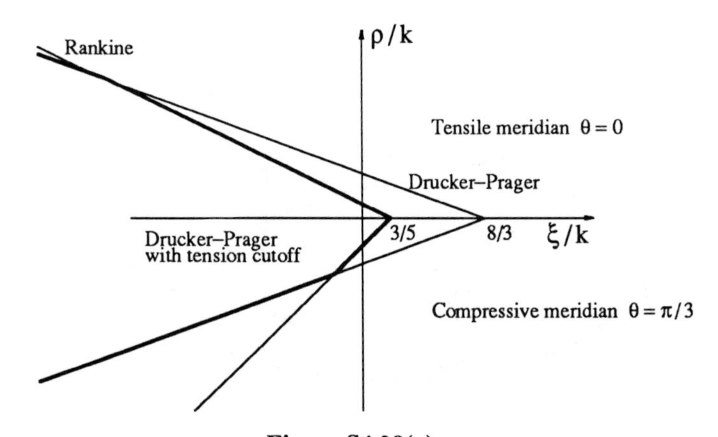

Figure S4.28(a).

The Drucker-Prager criterion in the meridian planes is reduced to

$$\frac{\rho}{k} = \sqrt{2}\,(1 - \frac{3}{8}\frac{\xi}{k})$$

The Rankine criterion in the meridian planes is reduced to

$$\frac{\rho_t}{k} = \frac{1}{\sqrt{2}}(\frac{3}{5} - \frac{\xi}{k}), \quad \frac{\rho_c}{k} = \sqrt{2}\,(\frac{3}{5} - \frac{\xi}{k})$$

They intersect at

tensile meridian: $\dfrac{\xi}{k} = -\dfrac{28}{5}, \quad \dfrac{\rho}{k} = \dfrac{31}{10}\sqrt{2}$

compressive meridian: $\dfrac{\xi}{k} = -\dfrac{16}{25}, \quad \dfrac{\rho}{k} = \dfrac{31}{25}\sqrt{2}$

(b) The combined surface in the (σ_1, σ_2) plane is shown in Fig. S4.28b.

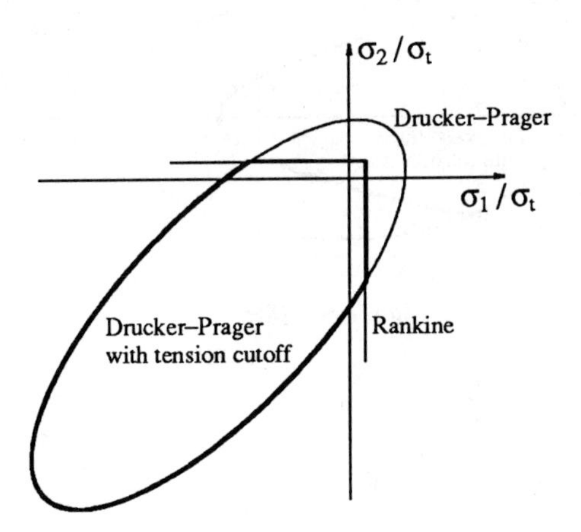

Figure S4.28(b).

The Drucker-Prager criterion in the (σ_1, σ_2) plane is reduced to

$$\left[\frac{x + \frac{12}{7}\sqrt{6}}{\frac{16}{7}\sqrt{6}}\right]^2 + \left[\frac{y}{4\sqrt{\frac{2}{7}}}\right]^2 = 1$$

where

$$x = \frac{1}{\sqrt{2}\,k}(\sigma_2 + \sigma_1), \quad y = \frac{1}{\sqrt{2}\,k}(\sigma_2 - \sigma_1)$$

The Rankine criterion in the (σ_1, σ_2) plane is reduced to

$$\frac{\sigma_1}{k} = \frac{\sqrt{3}}{5}, \quad \frac{\sigma_2}{k} = \frac{\sqrt{3}}{5}$$

They intersect at

$$\frac{\sigma_1}{k} = 0.3464, \quad \frac{\sigma_2}{k} = \pm 2.2559$$

(c) The surface in the (σ, τ) plane is shown in Fig. S4.28c.

The Drucker-Prager criterion in the (σ, τ) plane is reduced to

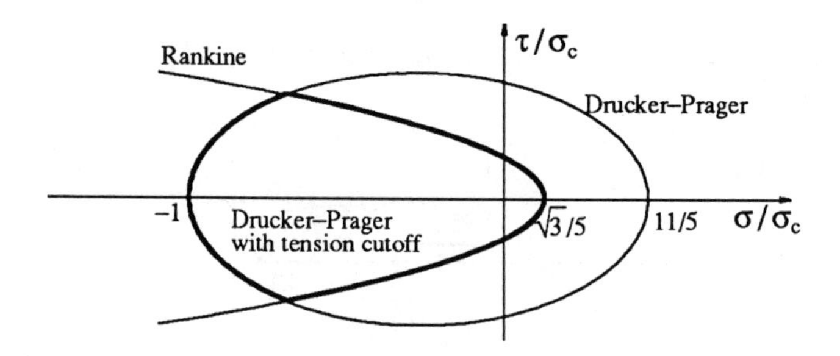

Figure S4.28(c).

$$\left[\frac{\sigma + \frac{24}{55}\sqrt{3}}{\frac{64}{55}\sqrt{3}\,k}\right]^2 + \left[\frac{\tau}{\frac{8}{\sqrt{55}}k}\right]^2 = 1$$

The Rankine criterion in (σ, τ) plane is reduced to

$$\frac{5}{\sqrt{3}}\frac{\sigma}{k} + \frac{25}{3}\left(\frac{\tau}{k}\right)^2 = 1$$

They intersect at

$$\frac{\sigma}{k} = -1.9106, \quad \frac{\tau}{k} = \pm 0.8842$$

Chapter 5

Perfectly Plastic Stress Analysis

5.1 General Aspects

For many practical applications, a material may be idealized as *perfectly plastic* that neglects the effect of strain-hardening. This idealization leads to a drastic simplification of the analysis of many complex structural engineering problems.

For the one-dimensional case, the elastic-perfectly plastic stress-strain relation can be expressed simply as, Equation (1.9),

$$\varepsilon = \frac{\sigma}{E} \qquad \sigma < \sigma_0$$

$$\varepsilon = \frac{\sigma}{E} + \lambda \qquad \sigma = \sigma_0$$

where σ_0 is the tensile yield strength of the material.

For the general case, the elastic-perfectly plastic stress-strain relationship can only be expressed in the incremental form as

$$d\sigma_{ij} = C_{ijkl}\, d\varepsilon_{ij}^{e} = C_{ijkl}\,(d\varepsilon_{ij} - d\varepsilon_{ij}^{p}) \tag{5.1}$$

where $d\sigma_{ij}$ is the stress increment tensor, C_{ijkl} the elastic stiffness tensor. $d\varepsilon_{ij}$, $d\varepsilon_{ij}^{e}$, and $d\varepsilon_{ij}^{p}$ are the total strain increment tensor, elastic strain increment tensor, and plastic strain increment tensor respectively, and they are related by

$$d\varepsilon_{ij} = d\varepsilon_{ij}^{e} + d\varepsilon_{ij}^{p} \tag{5.2}$$

Here, as in the one-dimensional case, the incremental strain is decomposed into two parts: an elastic component $d\varepsilon_{ij}^{e}$ and a plastic component $d\varepsilon_{ij}^{p}$.

For a given stress state at a point in a material, the yield condition or yield function $f(\sigma_{ij}) = 0$ determines whether the material is in an elastic state or a plastic state. If it is in a plastic state, we will then determine by the *loading criterion* whether a stress increment will constitute a plastic loading or an elastic unloading. If it is a plastic loading, then the direction of the corresponding plastic strain increment vector can be determined by the *flow rule.* Loading criteria, flow rules, and the development of the associated elastic-perfectly plastic stress-strain relationship are described in the following sections.

5.2 Loading Criterion

For a perfectly plastic material, the yield surface, $f(\sigma_{ij}) = 0$, is a fixed surface in stress space. Plastic deformation occurs if the current stress point, σ_{ij}, is on the surface, i.e., $f(\sigma_{ij}) = 0$. After the addition of the stress increment, $d\sigma_{ij}$, the resulting stress state, $\sigma_{ij} + d\sigma_{ij}$, must remain on the surface, or $f(\sigma_{ij} + d\sigma_{ij}) = 0$, in order to maintain the plastic flow. This is known as *loading.* On the other hand, if the resulting stress state moves inside the surface, or $f(\sigma_{ij} + d\sigma_{ij}) < 0$, no further plastic deformation occurs. This is called *unloading.* Thus, in terms of the stress tensor σ_{ij} and the stress increment tensor $d\sigma_{ij}$, the loading criterion is expressed as

$$\text{Loading:} \qquad f(\sigma_{ij}) = 0 \ \text{and} \ df = \frac{\partial f}{\partial \sigma_{ij}} d\sigma_{ij} = 0 \tag{5.3}$$

$$\text{Unloading:} \qquad f(\sigma_{ij}) = 0 \ \text{and} \ df = \frac{\partial f}{\partial \sigma_{ij}} d\sigma_{ij} < 0 \tag{5.4}$$

Since the yield function is also used here as a criterion for loading, the yield function is also called the *loading function.* Moreover, the condition $df = f(\sigma_{ij} + d\sigma_{ij}) - f(\sigma_{ij}) = 0$ is the *consistency* condition that is necessary for the determination of the magnitude of the plastic strain increment vector.

5.3 Plastic Potential and Flow Rule

The flow rule specifies the ratio of the components of the plastic strain increment tensor $d\varepsilon_{ij}^p$ or the direction of $d\varepsilon_{ij}^p$ in the strain space ε_{ij}.

A *plastic potential function* is often employed to describe a flow rule. A plastic potential function, $g(\sigma_{ij})$, is a scalar function of the stress tensor. The plastic strain increment vector corresponding to a given stress tensor σ_{ij} is specified as a vector normal to the potential function $g(\sigma_{ij})$ at σ_{ij},

$$d\varepsilon_{ij}^p = d\lambda \frac{\partial g}{\partial \sigma_{ij}} \tag{5.5}$$

where $d\lambda$ is a positive scalar and has non-zero value during a plastic loading.

The simplest case in the selection of a plastic potential function for an elastic-perfectly plastic material is to use the yield function as the potential function, i.e., $g = f$. Thus, Equation (5.5) becomes

$$d\varepsilon_{ij} = d\lambda \frac{\partial f}{\partial \sigma_{ij}} \tag{5.6}$$

This is generally referred to as *associated flow rule.* Equation (5.5), on the other hand, is referred to as *non-associated flow rule.* For materials with the associated flow rule, it can be proven that the solution of an elastic-plastic boundary problem is unique. In the following sub-sections, flow rules associated with various yield functions will be discussed.

5.3.1 Flow Rule Associated with Tresca Criterion

Let a coordinate system $(\varepsilon_1, \varepsilon_2, \varepsilon_3)$ in the principal strain space coincide with the coordinate system $(\sigma_1, \sigma_2, \sigma_3)$ in the principal stress space. Thus, the plastic strain increment $d\varepsilon_{ij}^p$ can be conveniently represented as a vector in the stress coordinate system $(\sigma_1, \sigma_2, \sigma_3)$

The Tresca yield surface is a right hexagonal prism consists of six planes in the principal stress space. Its meridian is parallel to the hydrostatic axis ξ. Thus, the projection of the plastic strain increment vector associated with the Tresca criterion on a meridian plane (ξ, ρ) is a vector perpendicular to the ξ axis and parallel to the ρ axis as shown in Fig. 5.1. This leads to

$$d\varepsilon_v^p = d\varepsilon_{ij}^p \delta_{ij} = 0 \tag{5.7}$$

that is, there is no plastic volumetric change. This characteristic holds for all hydrostatic-pressure-independent yield criteria such as the von Mises yield criterion.

On a deviatoric plane, the Tresca surface is a regular hexagon consisting of six planes and six singular edges. The directions of $d\varepsilon_{ij}^p$ for each plane and each edge must be determined separately. Figure 5.2 shows the plastic flow directions on each plane.

For example, for the case $\sigma_1 > \sigma_2 > \sigma_3$, the stress point is on the plane $\overline{AB}$

$$f(\sigma_{ij}) = \sigma_1 - \sigma_3 - \sigma_0 = 0 \tag{5.8}$$

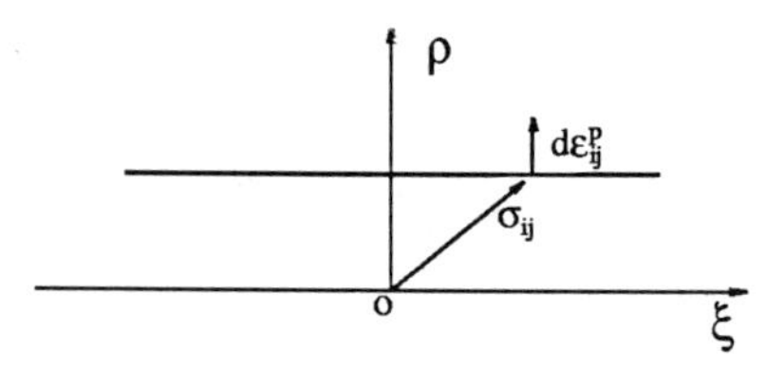

Figure 5.1. Flow rule associated with the Tresca criterion: meridian planes

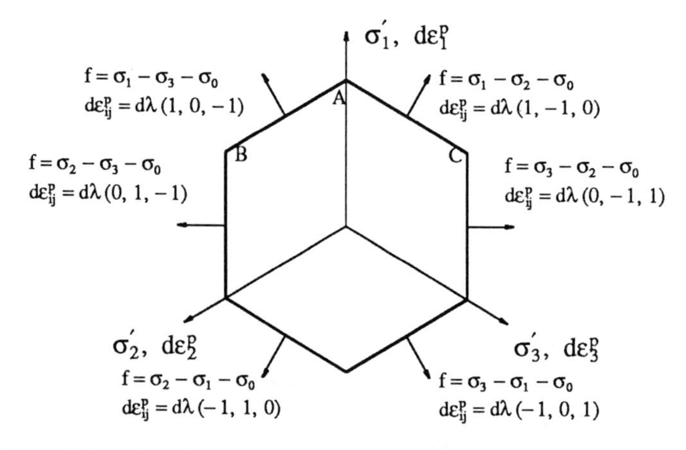

Figure 5.2. Flow rule associated with Tresca condition: deviatoric planes

and the plastic strain increment is expressed as

$$d\varepsilon_{ij}^p = (d\varepsilon_1^p,\ d\varepsilon_2^p,\ d\varepsilon_3^p) = d\lambda\,(1,\ 0,\ -1) \tag{5.9}$$

Along the edge of the hexagon, the situation is more complicated since the direction of plastic flow is undefined. For example, for the state at point A with $\sigma_1 \geq \sigma_2 = \sigma_3$ as shown in Fig. 5.2, the flow direction lies between of the flow direction normal to the plane $\overline{AB}$ and the flow direction normal to the plane $\overline{AC}$. The plastic strain increment vector at A is expressed as

$$d\varepsilon_{ij} = (d\varepsilon_1,\ d\varepsilon_2,\ d\varepsilon_3) = d\lambda\,(1,\ 0,\ -1) + d\mu\,(1,\ -1,\ 0) \tag{5.10}$$

where $d\lambda$ and $d\mu$ are both positive scalars.

5.3.2 Flow Rule Associated with the von Mises Criterion

The von Mises criterion has the form

$$f(\sigma_{ij}) = J_2 - k^2 = 0 \tag{5.11}$$

where $k^2 = \sigma_0^2/3$. Using the associated flow rule, Equation (5.6) becomes

$$d\varepsilon_{ij}^p = d\lambda \frac{\partial f}{\partial \sigma_{ij}} = d\lambda\, s_{ij} \tag{5.12}$$

that is, the plastic strain increment tensor is proportional to the deviatoric stress tensor s_{ij}. In component form, we write

$$\frac{d\varepsilon_x^p}{s_x} = \frac{d\varepsilon_y^p}{s_y} = \frac{d\varepsilon_z^p}{s_z} = \frac{d\gamma_{yz}^p}{2\,s_{yz}} = \frac{d\gamma_{zx}^p}{2\,s_{zx}} = \frac{d\gamma_{xy}^p}{2\,s_{xy}} = d\lambda \tag{5.13}$$

Equation (5.13) is known as the *Prandtl-Reuss equation.*

The projection of the plastic flow vector on a deviatoric plane is shown in Fig. 5.3. This direction is parallel to the direction of the projection of the stress vector or the deviatoric stress vector onto the same plane.

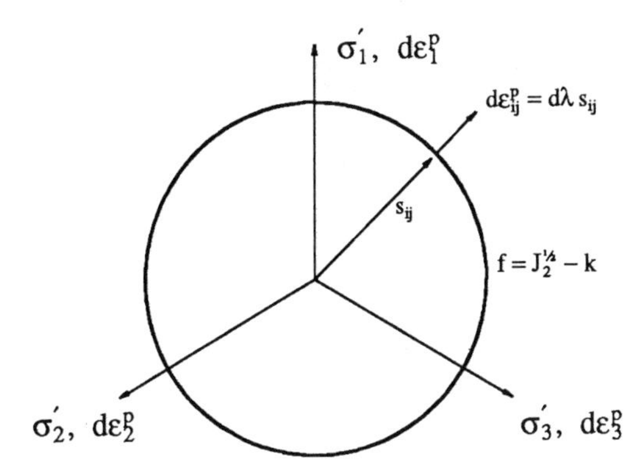

Figure 5.3. Flow rule associated with the von Mises criterion: deviatoric planes

5.3.3 Flow Rule Associated with the Rankine Criterion

The Rankine yield function has the form

$$\sigma_1 - \sigma_t = 0, \quad \text{if } \sigma_1 > \sigma_2 \text{ and } \sigma_1 > \sigma_3 \tag{5.14a}$$

$$\sigma_2 - \sigma_t = 0, \quad \text{if } \sigma_2 > \sigma_1 \text{ and } \sigma_2 > \sigma_3 \tag{5.14b}$$

$$\sigma_3 - \sigma_t = 0, \quad \text{if } \sigma_3 > \sigma_1 \text{ and } \sigma_3 > \sigma_2 \tag{5.14c}$$

The Rankine yield surface is a triangular prism in the principal stress space. The projections of plastic flow associated with the Rankine yield function in tensile and compressive meridian planes and in deviatoric planes are shown in Figs. 5.4 and 5.5 respectively. Note that the plastic volumetric strain increment is

$$d\varepsilon_v^p = d\varepsilon_{ij}^p \delta_{ij} = d\lambda \geq 0 \tag{5.15}$$

That is, the Rankine criterion always predicts a plastic volume expansion or dilatation. This is a common characteristic for elastic-plastic materials that are hydrostatic-pressure-dependent.

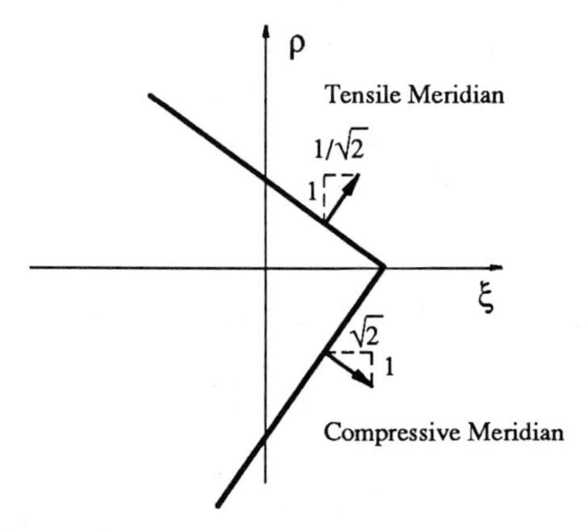

Figure 5.4. Flow rule associated with the Rankine criterion: meridian planes

5.3.4 Flow Rule Associated with the Mohr-Coulomb Criterion

The Mohr-Coulomb yield function has the simple form

$$m\,\sigma_{max} - \sigma_{min} = \sigma_c \tag{5.16}$$

where σ_c is the uniaxial compression yield strength. The Mohr-Coulomb yield surface is an irregular hexagonal prism in the principal stress space. The projections of plastic strain increment vector associated with the Mohr-Coulomb yield function in the tensile and compressive meridian planes and in deviatoric planes are shown in Figs. 5.6 and 5.7

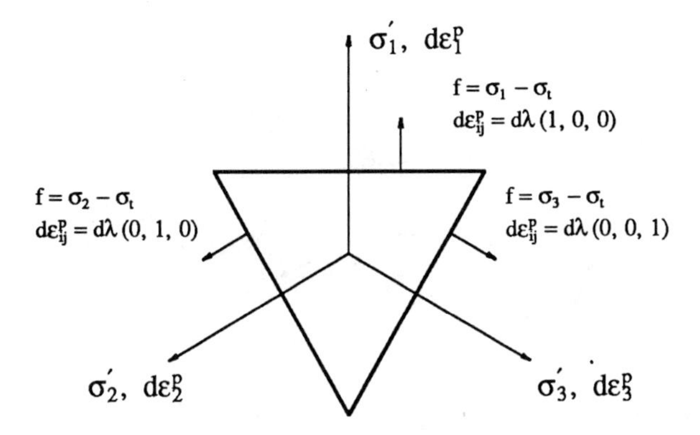

Figure 5.5. Flow rule associated with the Rankine criterion: deviatoric planes

respectively. The plastic volume dilatation predicted by the Mohr-Coulomb criterion has the value

$$d\varepsilon_v^p = d\varepsilon_{ij}^p \delta_{ij} = d\lambda\,(m-1) \tag{5.17}$$

Since $m > 1$ for the Mohr-Coulomb criterion, we have $d\varepsilon_v^p > 0$.

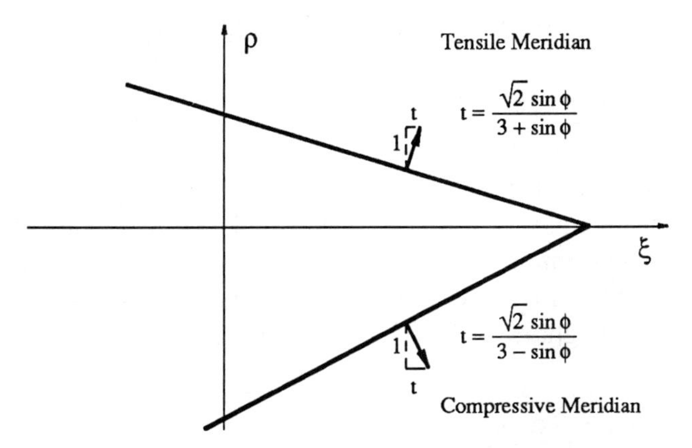

Figure 5.6. Flow rule associated with the Mohr-Coulomb criterion: meridian planes

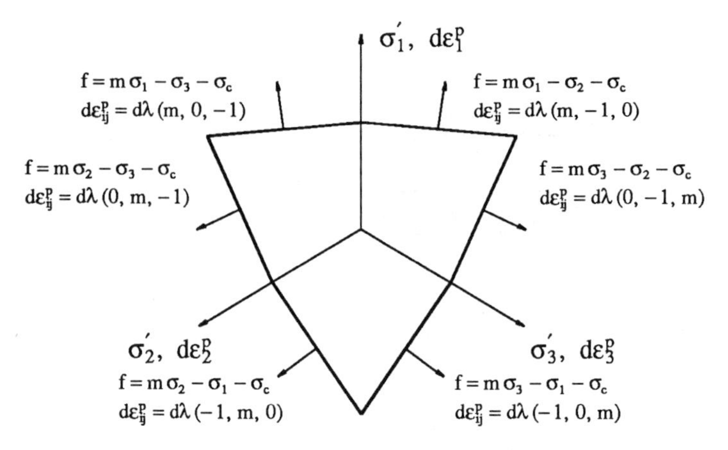

Figure 5.7. Flow rule associated with the Mohr-Coulomb criterion: deviatoric planes

5.3.5 Flow Rule Associated with the Drucker-Prager Criterion

The Drucker-Prager yield criterion has the form

$$\alpha I_1 + J_2^{1/2} - k = 0 \tag{5.18}$$

Using the associated flow rule, Equation (5.5) becomes

$$d\varepsilon_{ij}^p = \alpha\,\delta_{ij} + \frac{s_{ij}}{2\sqrt{J_2}} \tag{5.19}$$

The Drucker-Prager surface is a circular prism in the principal stress space. The projections of plastic flow associated with the Drucker-Prager yield function in the tensile and compressive meridian planes and in deviatoric planes are shown in Figs. 5.8 and 5.9 respectively. The plastic volume dilatation predicted by the Drucker-Prager criterion has the value

$$d\varepsilon_v^p = d\varepsilon_{ij}^p \delta_{ij} = 3\,\alpha\, d\lambda \tag{5.20}$$

Since $\alpha > 0$, we have $d\varepsilon_v^p > 0$.

5.4 Complete Perfectly Plastic Constitutive Relation

The complete incremental constitutive relationship expressing stress increment $d\sigma_{ij}$ in terms of total strain increment $d\varepsilon_{ij}$ is derived in the

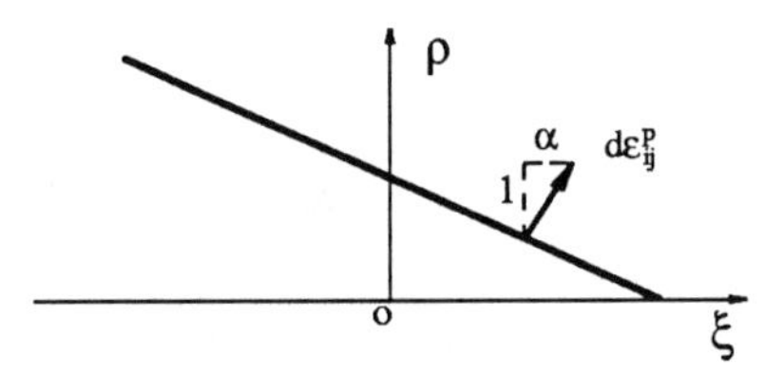

Figure 5.8. Flow rule associated with the Drucker-Prager criterion: meridian planes

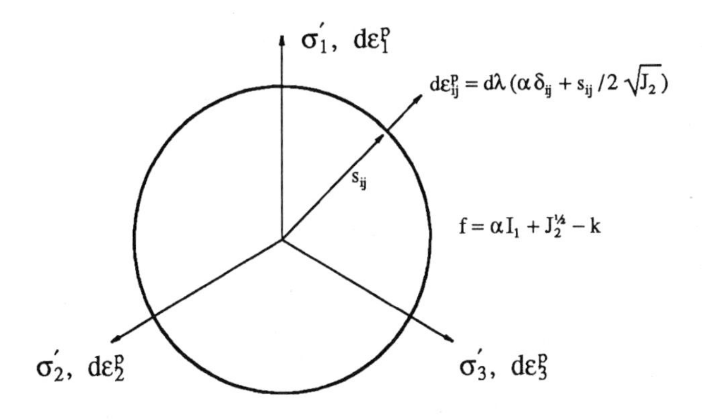

Figure 5.9. Flow rule associated with the Drucker-Prager criterion: deviatoric planes

following. Using the associated flow rule, Equation (5.1) has the form

$$d\sigma_{ij} = C_{ijkl}\,(d\varepsilon_{kl} - d\varepsilon_{kl}^p) = C_{ijkl}\,d\varepsilon_{kl} - d\lambda\,C_{ijkl}\,\frac{\partial f}{\partial \sigma_{kl}} \tag{5.21}$$

where C_{ijkl} is the elastic stiffness tensor. Combining Eq. (5.21) with the consistency condition

$$df = \frac{\partial f}{\partial \sigma_{ij}}\,d\sigma_{ij} = 0 \tag{5.22}$$

we obtain the positive scalar factor $d\lambda$

$$d\lambda = \frac{\dfrac{\partial f}{\partial \sigma_{ij}}\,C_{ijkl}}{\dfrac{\partial f}{\partial \sigma_{pq}}\,C_{pqrs}\,\dfrac{\partial f}{\partial \sigma_{rs}}}\,d\varepsilon_{kl} \tag{5.23}$$

Thus, the complete constitutive relationship for an elastic-perfectly plastic material has the form

$$d\sigma_{ij} = \left[C_{ijkl} - \frac{C_{ijmn} \dfrac{\partial f}{\partial \sigma_{mn}} \dfrac{\partial f}{\partial \sigma_{pq}} C_{pqkl}}{\dfrac{\partial f}{\partial \sigma_{rs}} C_{rstu} \dfrac{\partial f}{\partial \sigma_{tu}}} \right] d\varepsilon_{kl} = C^{ep}_{ijkl}\, d\varepsilon_{kl} \quad (5.24a)$$

$$\text{if } f(\sigma_{ij}) = 0 \text{ and } df = 0$$

$$d\sigma_{ij} = C_{ijkl}\, d\varepsilon_{kl} \quad (5.24b)$$

$$\text{if } f(\sigma_{ij}) = 0 \text{ and } df < 0 \quad \text{or } f(\sigma_{ij}) < 0$$

where C^{ep}_{ijkl} is the tangent moduli tensor for an elastic-perfectly plastic material.

5.5 Thin-walled Vessel Problems

Prob. 5.1 A long thin-walled circular tube with radius R and wall thickness t is subjected to an internal pressure p as shown in Fig. P5.1. Use the Tresca criterion with the following three end conditions: (1) the ends are free, (2) the ends are fixed, and (3) the ends are closed,

a. Express the Tresca criterion in terms of the pressure p;
b. Find the limit pressure $p = p_y$ at which the tube just yields; and
c. The ratio of the plastic strain increments when the tube just yields.

Solution: For all three cases, we have $\sigma_r = 0$ (thin-walled $t \ll R$). The only non-zero stress components are σ_θ and σ_z, and σ_θ and σ_z are principal stresses.

(1) Free Ends

From the equilibrium condition, we obtain $\sigma_\theta = \dfrac{pR}{t}$ and $\sigma_z = 0$. Thus, we have for this stress state, $\sigma_\theta \geq \sigma_z = \sigma_r = 0$. The Tresca criterion can be written as either $\sigma_\theta - \sigma_z = \sigma_0$ or $\sigma_\theta - \sigma_r = 0$. From either one of these two forms, we obtain

$$\sigma_\theta = \sigma_0\,, \quad \text{or} \quad \frac{pR}{t} = \sigma_0$$

The limit pressure is thus

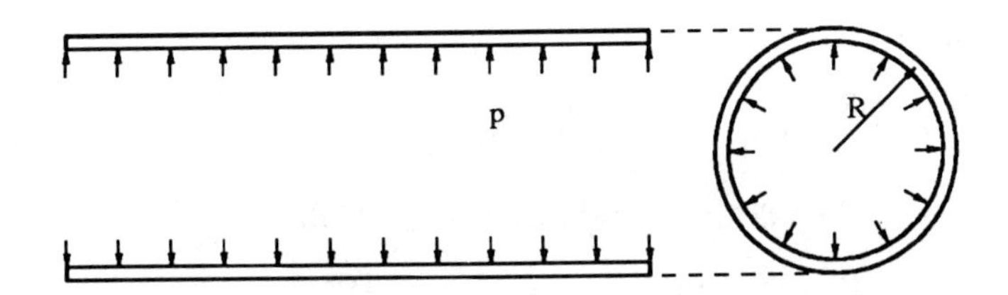

Figure P5.1.

$$p_y = \frac{t\sigma_0}{R}$$

If the yield condition $\sigma_\theta - \sigma_r = \sigma_0$ is used, the plastic strain increment is

$$d\varepsilon_{ij}^p = [\, d\varepsilon_\theta^p,\, d\varepsilon_z^p,\, d\varepsilon_r^p \,] = [1,\, 0,\, -1]\, d\lambda$$

while if $\sigma_\theta - \sigma_z = \sigma_0$ is used, the plastic strain increment is

$$d\varepsilon_{ij}^p = [\, d\varepsilon_\theta^p,\, d\varepsilon_z^p,\, d\varepsilon_r^p \,] = [1,\, -1,\, 0]\, d\lambda$$

(2) Fixed Ends

For this case, we have

$$\sigma_\theta = \frac{pR}{t}, \quad \sigma_z = \nu\,\frac{pR}{t}, \quad \sigma_r = 0, \qquad \sigma_\theta > \sigma_z > \sigma_r$$

Thus, the Tresca criterion is expressed as

$$\sigma_\theta - \sigma_r = \sigma_0, \quad \text{or} \quad \frac{pR}{t} = \sigma_0,$$

The limit pressure p_y and plastic strain increment are

$$p_y = \frac{t\sigma_0}{R}, \quad d\varepsilon_{ij}^p = [\, d\varepsilon_\theta^p,\, d\varepsilon_z^p,\, d\varepsilon_r^p \,] = [\,1,\, 0,\, -1\,]\, d\lambda$$

(3) Closed Ends

For this case, we have

$$\sigma_\theta = \frac{pR}{t}, \quad \sigma_z = \frac{pR}{2t}, \quad \sigma_r = 0, \qquad \sigma_\theta > \sigma_z > \sigma_r$$

The same limit pressure p_y and plastic strain increment as in Case (2) are found.

Prob. 5.2 Same thin-walled tube problem as in Prob. 5.1. Use the von Mises criterion.

Answer:

1. $\frac{pR}{t} = \sigma_0, \quad p_y = \frac{t\sigma_0}{R},$

   $d\varepsilon_{ij}^p = [\, d\varepsilon_\theta^p,\, d\varepsilon_z^p,\, d\varepsilon_r^p \,] = [\, 2, -1, -1 \,]\, d\lambda$

2. $\frac{pR}{t} = \frac{\sigma_0}{\sqrt{1 - \nu + \nu^2}}, \quad p_y = \frac{t\sigma_0}{R\sqrt{1 - \nu + \nu^2}},$

   $d\varepsilon_{ij}^p = [\, d\varepsilon_\theta^p,\, d\varepsilon_z^p,\, d\varepsilon_r^p \,] = [\, (2 - \nu), -(1 - 2\nu), -(1 + \nu) \,]\, d\lambda$

3. $\frac{pR}{t} = \frac{2}{\sqrt{3}}\sigma_0, \quad p_y = \frac{2t\sigma_0}{R\sqrt{3}},$

   $d\varepsilon_{ij}^p = [\, d\varepsilon_\theta^p,\, d\varepsilon_z^p,\, d\varepsilon_r^p \,] = [\, 1, 0, -1 \,]\, d\lambda$

Prob. 5.3 A long thin-walled tube with closed ends is subjected to an internal pressure p and torsional twist T at its ends. The tube has a radius R and wall thickness t. Denote the shear stress corresponding to the torsion twist T as τ, and $\sigma = \frac{pR}{2t}$. Using the Tresca criterion,

a. Express the Tresca criterion in terms of σ and τ;
b. For a constant internal pressure $\sigma = \sigma_0 / 4$, find the limit shear stress τ.

Answer:

a. $\tau^2 + 3\sigma_0\sigma - 2\sigma^2 = \sigma_0^2, \qquad \text{for } \sqrt{2}\,\sigma \geq \tau$

   $\sigma^2 + 4\tau^2 = \sigma_0^2, \qquad \text{for } \sqrt{2}\,\sigma < \tau$

b. $\tau = \frac{\sqrt{15}}{8}\sigma_0$

Prob. 5.4 Same thin-walled tube problem as in Prob. 5.3. Use the von Mises criterion.

Answer:

a. $\sigma^2 + \tau^2 = \frac{\sigma_0^2}{3} \qquad$ b. $\tau = \sqrt{\frac{13}{54}}$

Prob. 5.5 A long thin-walled circular tube with average diameter D and wall-thickness t is subjected to an internal pressure p_1 and an external pressure p_2, as shown in Fig. P5.5. Suppose that the external pressure p_2 does not contribute to the axial stress of the tube, and let $p_2 = rp_1$, $r \geq 0$. The tube yields under $p_1 = p_0$, and $p_2 = 0$.

a. Find the limit pressure $p_1 = p_y$, at which the tube just yields, in terms of p_0 and r for the case $r > 0$, according to i) von Mises criterion and ii) Tresca criterion;
b. Sketch the p_y vs. r curves for the two criteria; and
c. Find the values of r at which the limit pressures predicted by the two criteria have the largest difference, and give the reason.

Assume that the two criteria matched at the pure shear yield point.

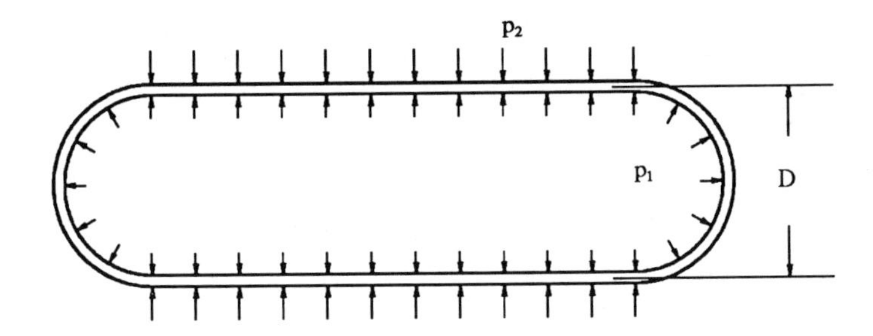

Figure P5.5.

Solution: Since t << D (thin-walled), the vessel is considered to be in a biaxial stress state, and the axial stress σ_a and the circumferential stress σ_c are given by

$$\sigma_a = \frac{D}{4t}\,p_1, \qquad \sigma_c = \frac{D}{2t}\,(p_1 - p_2) = \frac{D}{2t}\,(1-r)\,p_1$$

Since the von Mises and Tresca criterion are matched at the pure shear yield point, we shall first determine the shear yield stress k. For the case r = 0, the limit pressure is p_0, and we obtain the corresponding stresses as

$$\sigma_1 = \sigma_c = \frac{D}{2t}\,p_0, \quad \sigma_2 = \sigma_a = \frac{D}{4t}\,p_0, \quad \sigma_3 = 0$$

Using the Tresca criterion, we obtain the shear yield stress as

$$k = \frac{1}{2}\,(\sigma_1 - \sigma_3) = \frac{D}{4t}\,p_0$$

Thus, the two criteria can be expressed as

Tresca: $\frac{1}{2}(\sigma_1 - \sigma_3) = k$

von Mises: $\sigma_a^2 + \sigma_c^2 - \sigma_a\sigma_c = 3k^2$

Let $\bar{\sigma} = \frac{D}{4t} p_1$, we have

$$\sigma_a = \bar{\sigma}, \qquad \sigma_c = 2\bar{\sigma}(1 - r)$$

(a) Limit pressure p_y

von Mises criterion

Substituting σ_a and σ_c into von Mises criterion, we obtain

$$\bar{\sigma}^2 + 4\bar{\sigma}^2(1-r)^2 - 2\bar{\sigma}^2(1-r) = 3k^2$$

Solve for $\bar{\sigma}$,

$$\bar{\sigma} = \sqrt{\frac{3}{4r^2 - 6r + 3}}\, k$$

Thus, we obtain

$$p_1 = p_y = \sqrt{\frac{3}{4r^2 - 6r + 3}}\, p_0$$

Tresca criterion

For different values of r, the Tresca criterion has different forms, we must therefore consider each case separately.

(1) $0 \le r \le \frac{1}{2}$, in this case, we have

$$\sigma_1 = \sigma_c, \quad \sigma_2 = \sigma_a, \quad \sigma_3 = 0.$$

thus, the Tresca criterion becomes

$$\frac{1}{2}\sigma_c = k$$

or

$$\bar{\sigma}(1-r) = k, \qquad p_1 = p_y = \frac{p_0}{1-r}$$

(2) $\frac{1}{2} \le r \le 1$, in this case, we have

$$\sigma_1 = \sigma_a, \quad \sigma_2 = \sigma_c, \quad \sigma_3 = 0$$

and the Tresca criterion becomes

$$\frac{1}{2}\sigma_a = k$$

we obtain

$$p_y = 2p_0$$

(3) $1 \le r$, in this case,

$$\sigma_1 = \sigma_a, \quad \sigma_2 = 0, \quad \sigma_3 = \sigma_c < 0$$

and the Tresca criterion becomes

$$\frac{1}{2}(\sigma_a - \sigma_c) = k$$

or

$$\bar{\sigma}(2r-1) = 2k, \qquad p_y = \frac{p_0}{r - \frac{1}{2}}$$

Therefore, the limit pressure predicted by the Tresca criterion is

$$p_y = \begin{cases} \dfrac{p_0}{1-r}, & 0 \le r \le \dfrac{1}{2} \\ 2p_0, & \dfrac{1}{2} \le r \le 1 \\ \dfrac{p_0}{r - \frac{1}{2}}, & 1 \le r \end{cases}$$

(b) The p_y vs. r curves for the two criteria are plotted in Figure S5.5

(c) From Fig. S5.5, we see that the largest difference between the two limit pressures p_y as predicted by the two criteria occurs at $r = 1/2$ and $r = 1$.
At $r = 1/2$,

$$\sigma_a = \sigma_c = \bar{\sigma} > 0$$

and at $r = 1$,

$$\sigma_a = \bar{\sigma} > 0, \quad \sigma_c = 0.$$

In both cases, the material is yielded by tension. Since the two criteria are matched at the pure shear yield point, the two yield surfaces have the largest difference at the tension yield point.

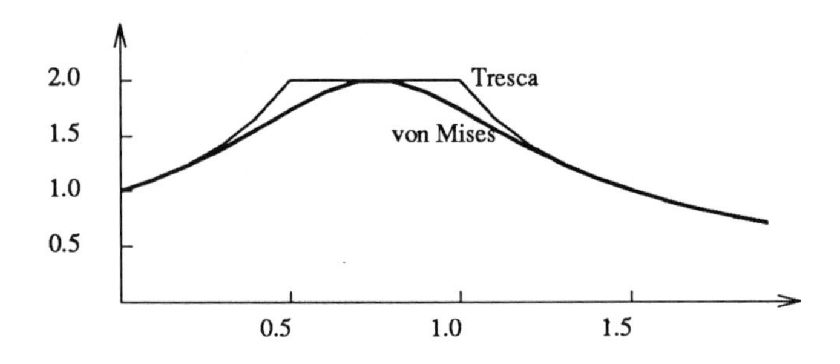

Figure S5.5.

Prob. 5.6 A thin-walled sphere of radius R and wall-thickness t is subjected to an internal pressure p. Using the von Mises criterion,

a. Express the von Mises criterion in terms of the pressure p;
b. Find the limit pressure p_y; and
c. Find the direction of the plastic flow when the sphere just yields.

Answer:

a. $\frac{pR}{2t} = \sigma_0$; b. $p_y = \frac{2\sigma_0 t}{R}$

c. $d\varepsilon_{ij}^p = [\, d\varepsilon_\theta^p,\ d\varepsilon_\phi^p,\ d\varepsilon_r^p \,] = [\,1,\ 1,\ -2\,]\, d\lambda$

Prob. 5.7 Same thin-walled sphere problem as in Prob. 5.6. Use the Tresca criterion.

Answer:

a. $\frac{pR}{2t} = \sigma_0$; b. $p_y = \frac{2\sigma_0 t}{R}$

c. $d\varepsilon_{ij}^p = [\, d\varepsilon_\theta^p,\ d\varepsilon_\phi^p,\ d\varepsilon_r^p \,] = [\,1,\ 0,\ -1\,]\, d\lambda + [\,0,\ 1,\ -1\,]\, d\mu$

Prob. 5.8 In a combined tension/torsion test on a thin-wall tube of circular cross-section, let σ, ε be the axial stress and axial strain and τ, γ be the shear stress and shear strain, respectively. Assume that the tube is made of Prandtl-Reuss material with $\nu = 1/2$. Calculate the stresses σ and τ corresponding to the strain state $(\varepsilon, \gamma) = (\sigma_0 / E\ ,\ \sigma_0 / \sqrt{3}G)$ for the following three loading paths, as shown in Fig. P5.8

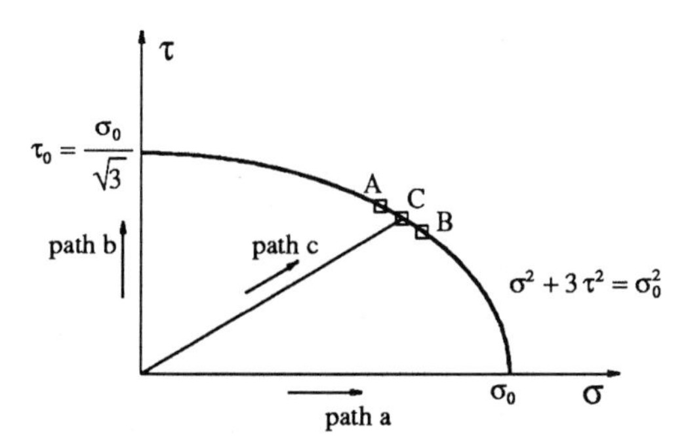

Figure S5.8.

a. The axial strain ε is first increased up to the yield value: $\varepsilon = \sigma_0 / E$, then kept unchanged, while the shear strain γ is increased up to its final value $\gamma = \sigma_0 / \sqrt{3}\,G$;
b. Reverse the above loading path: the shear strain γ is first increased up to its final value $\gamma = \sigma_0 / \sqrt{3}\,G$, then held constant, while the axial strain ε is increased up to its final value σ_0 / E.
c. Both strains ε and γ are proportionally increased with a ratio of $\varepsilon / \gamma = \sqrt{3}\,G / E = 1 / \sqrt{3}$, until their final values are reached.

Solution: The yield function of the Prandtl-Reuss material for the given stress state is

$$f = \sigma^2 + 3\tau^2 - \sigma_0^2 = 0 \tag{1}$$

and the incremental stress-strain relation is

$$d\sigma_{ij} = 2G\, de_{ij} + K\, d\varepsilon_{kk}\, \delta_{ij} - \frac{G\, s_{mn}\, de_{mn}}{k^2}\, s_{ij} \tag{2}$$

Using the incompressibility condition, we obtain

$$d\varepsilon_{kk} = 0, \;\; de_{ij} = d\varepsilon_{ij}, \;\; G = E/3$$

and

$$s_{mn}\, de_{mn} = \sigma\, d\varepsilon + \tau\, d\gamma$$

Therefore, in component form, the incremental stress-strain relation becomes

$$d\sigma = E(1 - \frac{\sigma^2}{3k^2})\, d\varepsilon - \frac{E}{3k^2}\, \sigma\tau\, d\gamma \tag{3}$$

$$d\tau = - \frac{E}{3k^2}\, \sigma\tau\, d\varepsilon + \frac{E}{3}\, (1 - \frac{\tau^2}{k^2})\, d\gamma \tag{4}$$

For the elastic loading, we have

$$d\sigma = E\, d\varepsilon, \quad d\tau = \frac{E}{3}\, d\gamma \tag{5}$$

Path a: $(\varepsilon, \gamma) = (0, 0) \rightarrow (\frac{\sigma_0}{E}, 0) \rightarrow (\frac{\sigma_0}{E}, \frac{\sigma_0}{\sqrt{3}\, G})$

In the path that ε increases form 0 to $\frac{\sigma_0}{E}$ and $\gamma = 0$, the material is in an elastic loading state, and at the end of the path, $(\sigma, \tau) = (\sigma_{0,}\ 0)$, the material begins to yield.

For the path that γ increases from 0 to $\frac{\sigma_0}{\sqrt{3G}}$ while ε remains constant, the material is in a plastic loading state with $d\varepsilon = 0$, and we have

$$d\sigma = -\frac{E}{3k^2}\, \sigma\tau\, d\gamma \tag{6}$$

$$d\tau = \frac{E}{3}\, (1 - \frac{\tau^2}{k^2})\, d\gamma \tag{7}$$

We shall first solve the ordinary differential equation, Eq. (7). Rewriting Equation (7) in the form

$$\frac{d\tau}{1 - \tau^2/k^2} = \frac{E}{3}\, d\gamma$$

and integrating the above equation, we obtain

$$\frac{E}{3}\, (\gamma - \gamma_0) = \frac{k}{2}\, ln\, \frac{1 + \tau/k}{1 - \tau/k} + C$$

Using the condition, at $\gamma = \gamma_0 = 0$, $\tau = 0$, and $k = \frac{\sigma_0}{\sqrt{3}}$, and rearranging the result, we obtain the shear stress τ at the end of the path

$$\tau = \frac{1}{\sqrt{3}}\, \frac{e^2 - 1}{e^2 + 1}\, \sigma_0 = 0.4397\, \sigma_0$$

Using yield condition, Eq. (1), we obtain the normal stress σ at the end of the path

$$\sigma = \sqrt{\sigma_0^2 - 3\tau^2} = 0.6481\, \sigma_0$$

Path b: $(\varepsilon, \gamma) = (0, 0) \rightarrow (0, \frac{\sigma_0}{\sqrt{3}\, G}) \rightarrow (\frac{\sigma_0}{E}, \frac{\sigma_0}{\sqrt{3}\, G})$

For the path that $\varepsilon = 0$ and γ increases from 0 to $\frac{\sigma_0}{\sqrt{3}\, G}$, the material is in an elastic loading state, and at the end of the path, $(\sigma, \tau) = (0, \frac{\sigma_0}{\sqrt{3}})$, the material begins to yield. For the path that ε increases from 0 to $\frac{\sigma_0}{E}$ while γ remains constant, the material is in a plastic loading state with $d\gamma = 0$, and we have

$$d\sigma = E(1 - \frac{\sigma^2}{3k^2})\, d\varepsilon \tag{8}$$

$$d\tau = -\frac{E}{3k^2}\, \sigma\tau\, d\varepsilon \tag{9}$$

Solving Equation (8) for σ, we obtain

$$E\, d\varepsilon = \frac{1}{2}\, \frac{d\sigma}{1 + \sigma/\sqrt{3}k} + \frac{1}{2}\, \frac{d\sigma}{1 - \sigma/\sqrt{3}k}$$

Integrating the above equation, we obtain

$$E(\varepsilon - \varepsilon_0) = \frac{\sqrt{3}k}{2}\, ln\, \frac{1 + \sigma/\sqrt{3}k}{1 - \sigma/\sqrt{3}k} + C$$

Using the condition, at $\varepsilon = \varepsilon_0 = 0$, $\sigma = 0$, and $k = \frac{\sigma_0}{\sqrt{3}}$, we obtain the normal stress σ at the end of the path

$$\sigma = \frac{e^2-1}{e^2+1}\, \sigma_0 = 0.7616\, \sigma_Y$$

Substituting the above equation to the yield criterion, we obtain the shear stress τ at the end of the path

$$\tau = \frac{1}{\sqrt{3}}\, \sqrt{\sigma_0^2 - \sigma^2} = 0.3742\, \sigma_0$$

Path c: $(\varepsilon, \gamma) = (0, 0) \rightarrow (\frac{\sigma_0}{E}, \frac{\sigma_0}{\sqrt{3}\, G})$

This path is a proportional loading path. Denoting the proportional factor r, $0 \le r \le 1$, such that at any point on the path, the strains can be expressed as

$$\varepsilon = r\, \frac{\sigma_0}{E}, \qquad \gamma = r\, \frac{\sigma_0}{\sqrt{3}\, G} = r\, \frac{\sqrt{3}\sigma_0}{E} \tag{10}$$

At a point on the path, $r = r^*$, the material yields. Using the yield condition, Eq. (1), we obtain r^* and the stresses at the point as

$$r^* = \frac{1}{\sqrt{2}}, \quad \sigma^* = \frac{1}{\sqrt{2}}\, \sigma_0, \quad \tau^* = \frac{1}{\sqrt{2}}\, \frac{\sigma_0}{\sqrt{3}} = \frac{\sigma_0}{\sqrt{6}} \tag{11}$$

For the path that r increases from r^* to 1, the material is in plastic loading state. Substituting Eq. (10) to Eqs. (3) and (4), we have

$$d\sigma = [(1 - \frac{\sigma^2}{3k^2})\,\sigma_0 - \frac{\sqrt{3}\,\sigma_0}{3k^2}\,\sigma\tau]\,dr \tag{12}$$

$$d\tau = [(1 - \frac{\tau^2}{k^2})\,\frac{\sigma_0}{\sqrt{3}} - \frac{\sigma_0}{3k^2}\,\sigma\tau]\,dr \tag{13}$$

Using the yield condition, Eq. (1), Eqs. (12) and (13) can be rewritten as

$$\sigma_0\,\frac{d\sigma}{dr} = \sqrt{3}\,\tau\,(\sqrt{3}\,\tau - \sigma) \tag{14}$$

$$\sigma_0\,\frac{d\tau}{dr} = -\,\frac{\sigma}{\sqrt{3}}\,(\sqrt{3}\,\tau - \sigma) \tag{15}$$

From the above two equations, we obtain

$$\frac{d\sigma}{d\tau} = -\,\frac{3\tau}{\sigma}$$

This implies that the stress point will move in the tangential direction of the yield surface. At the point (σ^*, τ^*), Equations (10), (14) and (15) leads to

$$\frac{d\sigma}{dr} = \frac{\sigma_0}{E}\,\frac{d\sigma}{d\varepsilon} = \frac{\sqrt{3}\,\sigma_0}{E}\,\frac{d\sigma}{d\gamma} = 0, \qquad \frac{d\tau}{dr} = \frac{\sigma_0}{E}\,\frac{d\tau}{d\varepsilon} = \frac{\sqrt{3}\,\sigma_0}{E}\,\frac{d\tau}{d\gamma} = 0$$

This implies that the stress point can not move any further from the point (σ^*, τ^*). Therefore, we have, at the end of the path

$$\sigma = \sigma^* = \frac{1}{\sqrt{2}}\,\sigma_0 = 0.7071\,\sigma_0, \quad \tau = \tau^* = \frac{\sigma_0}{\sqrt{6}} = 0.4082\,\sigma_0$$

5.6 Plastic Zone Near Crack Tip Problems

The near tip stress field of a crack derived from the linear elastic fracture mechanics has a stress field with singularity: the stress components tend to infinite as the distance to the tip approaches to zero. There therefore exists a plastic zone near the crack tip. Elastic-plastic analysis is necessary to obtain a realistic solution including the near tip stress field and the spreading shape of the plastic zone.

It is very difficult to find closed form solutions for crack problems in the elastic-plastic fracture analysis, and only a few solutions are available for very simple cases. For these simple problems, we may substitute the stress components obtained from the elastic analysis into a yield criterion to obtain the approximate shape of the plastic zone near the tip. This approach is obviously not rigorous. Although the size of the plastic zone so obtained is usually smaller than the actual one, it does provide a reasonable qualitative understanding of the spreading of yielding for crack problems involving small scale yielding.

Herein, the spread of plasticity near the tip for the mode I and mode II cracks as well as for the mixed mode combining both mode I and mode II cracks are studied. To this end, the following five yield criteria are used: von-Mises, Tresca, Rankine, Mohr-Coulomb, and Drucker-Prager criterion.

The general expression for the spread of plasticity for the mixed mode crack is difficult to derive and recourse must be made to a numerical procedure. A numerical procedure has been implemented in the program PLASTIC_ZONE. The listing of the source code for PLASTIC_ZONE is given in Section 5.9. Figure 5.10 shows the shapes of the spread of plasticity for mode I, mode II and mixed mode crack under plane strain condition and using the Tresca yield criterion.

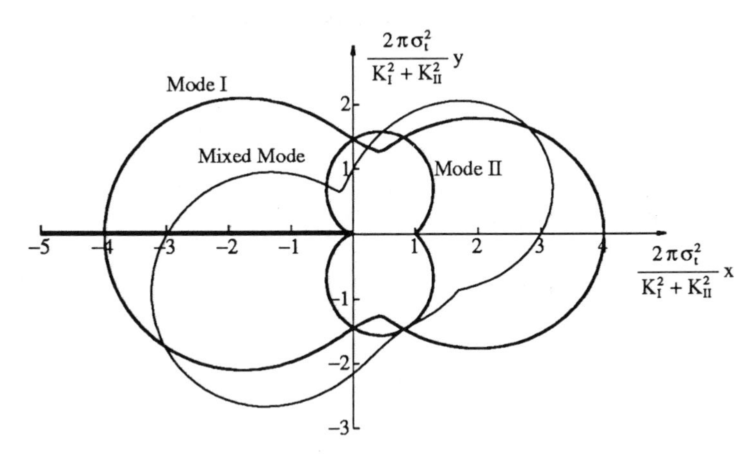

Figure 5.10. Spread of plasticity near crack tip calculated by PLASTIC_ZONE

In the following, the elastic stress field for a general mixed crack is first reviewed, followed by the spread of plasticity solutions with various yield criteria.

An infinite plane with a line crack of length 2a subjected to a biaxial tension stress σ_0 and a shear stress τ_0 at infinity is shown in Fig. 5.11. The origin of both coordinate systems (x, y) and (r, θ) is located at the crack tip. Thus, the near tip stress field for a general mixed mode crack can be written as

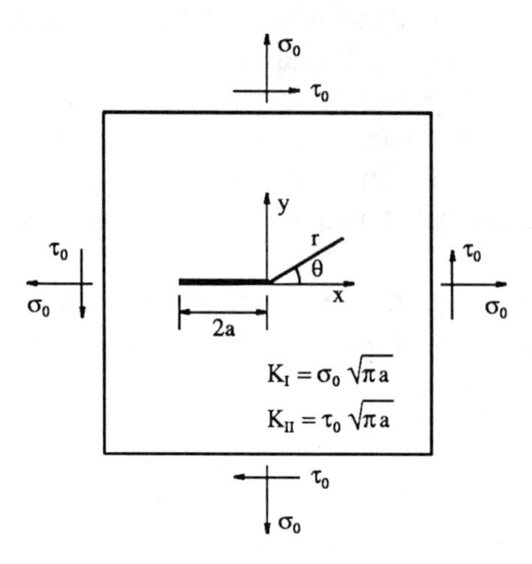

Figure 5.11. A general mixed-mode crack

$$\sigma_{ij} = \frac{K_I}{\sqrt{2\pi r}} f_{ij}(\theta) + \frac{K_{II}}{\sqrt{2\pi r}} g_{ij}(\theta), \quad (i,j=1,2) \tag{5.25}$$

where K_I and K_{II} are the stress intensity factors for mode I and mode II respectively. The functions $f_{ij}(\theta)$ and $g_{ij}(\theta)$ have the forms

$$f_{11}(\theta) = f_{xx}(\theta) = \cos\frac{\theta}{2}\,[\,1 - \sin\frac{\theta}{2}\sin\frac{3\theta}{2}\,] \tag{5.26a}$$

$$f_{22}(\theta) = f_{yy}(\theta) = \cos\frac{\theta}{2}\,[\,1 + \sin\frac{\theta}{2}\sin\frac{3\theta}{2}\,] \tag{5.26b}$$

$$f_{12}(\theta) = f_{21}(\theta) = f_{xy}(\theta) = \cos\frac{\theta}{2}\sin\frac{\theta}{2}\cos\frac{3\theta}{2} \tag{5.26c}$$

$$g_{11}(\theta) = g_{xx}(\theta) = -\sin\frac{\theta}{2}\,[\,2 + \cos\frac{\theta}{2}\cos\frac{3\theta}{2}\,] \tag{5.27a}$$

$$g_{22}(\theta) = g_{yy}(\theta) = \sin\frac{\theta}{2}\cos\frac{\theta}{2}\cos\frac{3\theta}{2} \tag{5.27b}$$

$$g_{12}(\theta) = g_{21}(\theta) = g_{xy}(\theta) = \cos\frac{\theta}{2}\,[\,1 - \sin\frac{\theta}{2}\sin\frac{3\theta}{2}\,] \tag{5.27c}$$

where $-\pi \leq \theta \leq \pi$. Denote

$$\bar{K} = \sqrt{K_I^2 + K_{II}^2}, \quad k_1 = \frac{K_I}{\bar{K}}, \quad k_2 = \frac{K_{II}}{\bar{K}} \tag{5.28}$$

we can write

$$\sigma_{ij} = \frac{\bar{K}}{\sqrt{2\pi r}} \bar{\sigma}_{ij} \tag{5.29}$$

or

$$\sigma_{xx} = \frac{\bar{K}}{\sqrt{2\pi r}} \bar{\sigma}_{xx} = \frac{\bar{K}}{\sqrt{2\pi r}} [\, k_1 \, f_{xx}(\theta) + k_2 \, g_{xx}(\theta)] \tag{5.30a}$$

$$\sigma_{yy} = \frac{\bar{K}}{\sqrt{2\pi r}} \bar{\sigma}_{yy} = \frac{\bar{K}}{\sqrt{2\pi r}} [\, k_1 \, f_{yy}(\theta) + k_2 \, g_{yy}(\theta)] \tag{5.30b}$$

$$\sigma_{xy} = \frac{\bar{K}}{\sqrt{2\pi r}} \bar{\sigma}_{xy} = \frac{\bar{K}}{\sqrt{2\pi r}} [\, k_1 \, f_{xy}(\theta) + k_2 \, g_{xy}(\theta)] \tag{5.30c}$$

and

$$\sigma_{zz} = \begin{cases} 0 & \text{for plane stress} \\ \dfrac{\bar{K}}{\sqrt{2\pi r}} \nu \, (\bar{\sigma}_{xx} + \bar{\sigma}_{yy}) & \text{for plane strain} \end{cases} \tag{5.30d}$$

where ν is Poisson's ratio.

The five yield criteria under consideration can all be expressed generally as

$$f(\sigma_{ij}) - \sigma_t = 0 \tag{5.31}$$

where σ_t is the tensile yield stress, and $f(\sigma_{ij})$ is a homogeneous function of degree one in stresses. Using the stress tensor components, we can write

$$\frac{\bar{K}}{\sqrt{2\pi r}} f(\bar{\sigma}_{ij}) - \sigma_t = 0 \tag{5.32}$$

Define

$$\bar{k} = \frac{\bar{K}}{\sigma_t} \tag{5.33}$$

and we solve the radius of the plastic zone from the above equations as

$$r_p = \frac{\bar{k}^2}{2\pi} f^2(\bar{\sigma}_{ij}) = \frac{\bar{k}^2}{2\pi} R(\theta) = \frac{\bar{K}_I^2 + \bar{K}_{II}^2}{2\pi\sigma_t^2} R(\theta) \tag{5.34}$$

$$R(\theta) = f^2(\bar{\sigma}_{ij}) \tag{5.35}$$

For the crack tip with the x-y coordinate system defined in Fig. 5.11, the shape of the plastic zone can be expressed as

$$\bar{x} = \frac{2\pi\sigma_t^2}{K_I^2 + K_{II}^2} x = R(\theta) \cos\theta \tag{5.36a}$$

$$\bar{y} = \frac{2\pi\sigma_t^2}{K_I^2 + K_{II}^2} y = R(\theta) \sin\theta \tag{5.36b}$$

Prob. 5.9 Show that the invariants of the stress field near mode I crack under plane stress are given below. Plot principal stresses vs. θ curves.

$$\sigma_1 = \frac{\bar{K}}{\sqrt{2\pi r}} \bar{\sigma}_1, \quad \bar{\sigma}_1 = \cos\frac{\theta}{2}\left(1 + \left|\sin\frac{\theta}{2}\right|\right)$$

$$\sigma_2 = \frac{\bar{K}}{\sqrt{2\pi r}} \bar{\sigma}_2, \quad \bar{\sigma}_2 = \cos\frac{\theta}{2}\left(1 - \left|\sin\frac{\theta}{2}\right|\right)$$

$$\sigma_3 = \frac{\bar{K}}{\sqrt{2\pi r}} \bar{\sigma}_3, \quad \bar{\sigma}_3 = 0$$

$$\bar{\sigma}_1 \geq \bar{\sigma}_2 \geq \bar{\sigma}_3, \quad \text{for } -\pi \leq \theta \leq \pi$$

$$I_1 = \frac{\bar{K}}{\sqrt{2\pi r}} \bar{I}_1, \quad \bar{I}_1 = 2\cos\frac{\theta}{2}$$

$$J_2 = \frac{\bar{K}}{\sqrt{2\pi r}} \bar{J}_2, \quad \bar{J}_2 = \frac{1}{3}\left[1 + 3\sin^2\frac{\theta}{2}\right]\cos^2\frac{\theta}{2}$$

Answer: Figure S5.9.

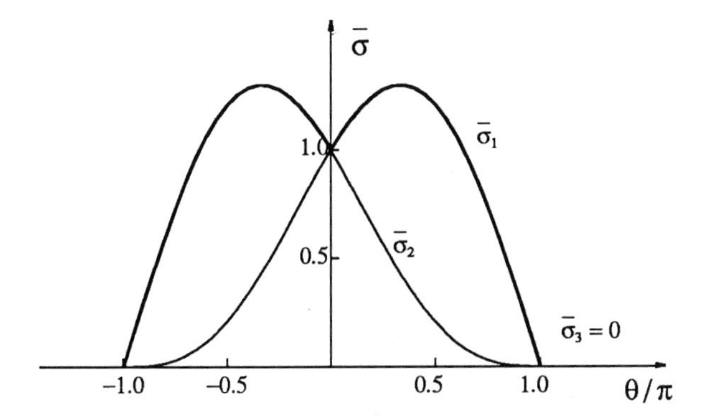

Figure S5.9. Principal stresses of Mode I under plane stress condition

Prob. 5.10 Show that the invariants of the stress field near mode I crack under plane strain are given below. Plot principal stresses vs. θ curves.

$$\bar{\sigma}_1 = \cos\frac{\theta}{2}\,(1 + |\sin\frac{\theta}{2}|)$$

$$\bar{\sigma}_2 = \cos\frac{\theta}{2}\,(1 - |\sin\frac{\theta}{2}|)$$

$$\bar{\sigma}_3 = 2\,\nu\cos\frac{\theta}{2}$$

$$\bar{\sigma}_1 \geq \bar{\sigma}_3 \geq \bar{\sigma}_2, \quad \text{for } -\pi \leq \theta \leq -\theta_I$$

$$\bar{\sigma}_1 \geq \bar{\sigma}_2 \geq \bar{\sigma}_3, \quad \text{for } -\theta_I \leq \theta \leq \theta_I$$

$$\bar{\sigma}_1 \geq \bar{\sigma}_3 \geq \bar{\sigma}_2, \quad \text{for } \theta_I \leq \theta \leq \pi$$

where

$$\theta_I = 2\sin^{-1}(1 - 2\nu)$$

$$\bar{I}_1 = 2\,(1 + \nu)\cos\frac{\theta}{2}$$

$$\bar{J}_2 = \frac{1}{3}\,[(1 - 2\nu^2) + 3\sin^2\frac{\theta}{2}]\cos^2\frac{\theta}{2}$$

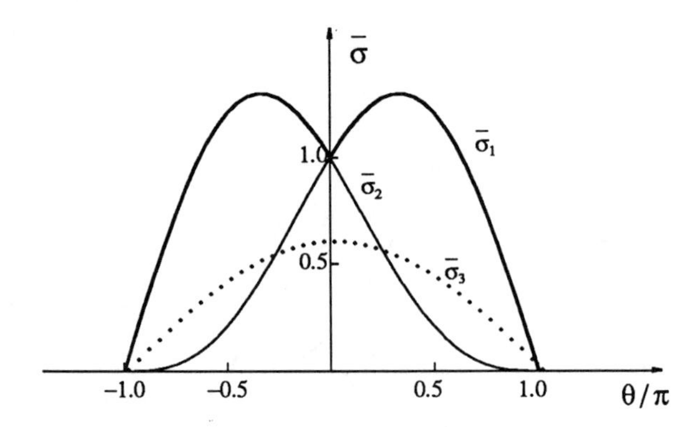

Figure S5.10. Principal stresses of Mode I under plane strain condition

Answer: Figure S5.10.

Prob. 5.11 Show that the invariants of the stress field near mode II crack under plane stress condition are given below. Plot principal stresses vs. θ curves.

$$\bar{\sigma}_1 = -\sin\frac{\theta}{2} + \frac{1}{2}\sqrt{4 - 3\sin^2\theta}$$

$$\bar{\sigma}_2 = -\sin\frac{\theta}{2} - \frac{1}{2}\sqrt{4 - 3\sin^2\theta}$$

$$\bar{\sigma}_3 = 0$$

$$\bar{\sigma}_1 \geq \bar{\sigma}_2 \geq \sigma_3 \quad \text{for} \quad -\pi \leq \theta \leq -\theta_{\text{II}}$$

$$\bar{\sigma}_1 \geq \bar{\sigma}_3 \geq \sigma_2 \quad \text{for} \ -\theta_{\text{II}} \leq \theta \leq \theta_{\text{II}}$$

$$\bar{\sigma}_3 \geq \bar{\sigma}_1 \geq \sigma_2 \quad \text{for} \quad \theta_{\text{II}} \leq \theta \leq \pi$$

where

$$\theta_{\text{II}} = 2\sin^{-1}\frac{1}{\sqrt{3}}$$

$$\bar{I}_1 = -2\sin\frac{\theta}{2}$$

$$\bar{J}_2 = \frac{1}{3}\,[3 - 8\sin^2\frac{\theta}{2} + 9\sin^4\frac{\theta}{2}]$$

Answer: Figure S5.11.

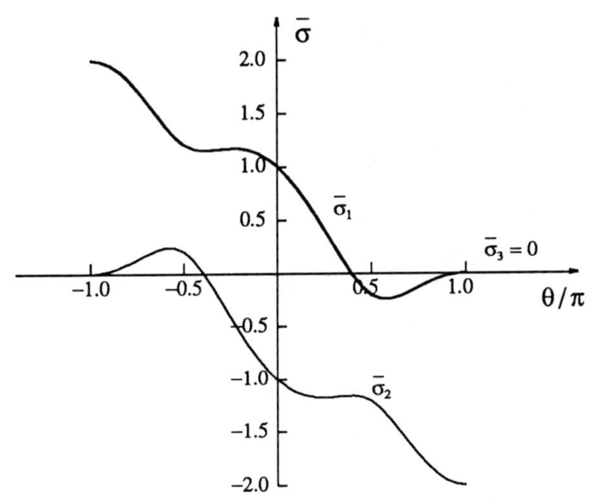

Figure S5.11. Principal stresses of Mode II in plane stress condition

Prob. 5.12 Show that the invariants of the stress field near mode II crack under plane strain condition are given below. Plot principal stresses vs. θ curves.

$$\bar{\sigma}_1 = -\sin\frac{\theta}{2} + \frac{1}{2}\sqrt{4 - 3\sin^2\theta}$$

$$\bar{\sigma}_2 = -\sin\frac{\theta}{2} + \frac{1}{2}\sqrt{4 - 3\sin^2\theta}$$

$$\bar{\sigma}_3 = -2\nu\sin\frac{\theta}{2}$$

$$\bar{\sigma}_1 \geq \bar{\sigma}_3 \geq \bar{\sigma}_2, \quad \text{for} -\pi \leq \theta \leq \pi$$

$$\bar{I}_1 = -2\,(1 + \nu)\sin\frac{\theta}{2}$$

$$\bar{J}_2 = \frac{1}{3}[3 - (8 + 4\nu - 4\nu^2)\sin^2\frac{\theta}{2} + 9\sin^4\frac{\theta}{2}]$$

Answer: Figure S5.12.

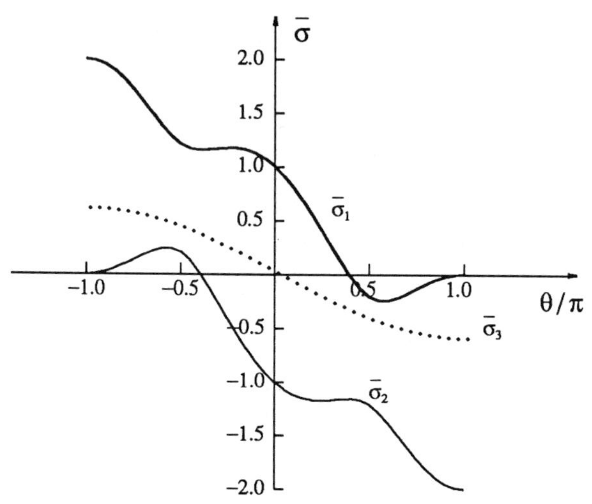

Figure S5.12. Principal stresses of Mode II under plane strain condition

Prob. 5.13 Derive the expressions of the shapes of plastic zone near mode I crack and plot the shapes in the (x, y) plane. Use the Tresca criterion for both plane stress and plane strain cases.

Solution: The Tresca criterion can be written as

$$\sigma_{max} - \sigma_{min} = \sigma_t$$

where σ_{max} and σ_{min} are the maximum and minimum principal stresses respectively. Using Equations (5.31) and (5.35), we have

$$R(\theta) = (\bar{\sigma}_{max} - \bar{\sigma}_{min})^2$$

Substituting the stress expression in Probs. 5.9 and 5.10, the shapes of plastic zone for mode I crack can be obtained as

Plane Stress:

$$R(\theta) = \begin{cases} \cos^2\frac{\theta}{2}\,(1 - \sin\frac{\theta}{2})^2 & -\pi \le \theta \le 0 \\ \cos^2\frac{\theta}{2}\,(1 + \sin\frac{\theta}{2})^2 & 0 \le \theta \le \pi \end{cases}$$

Plane Strain:

$$R(\theta) = \begin{cases} \sin^2\theta & -\pi \le \theta \le -\theta_I \\ \cos^2\theta\,[(1-2\nu) - \sin\dfrac{\theta}{2}]^2 & -\theta_I \le \theta \le 0 \\ \cos^2\theta\,[(1-2\nu) + \sin\dfrac{\theta}{2}]^2 & 0 \le \theta \le \theta_I \\ \sin^2\theta & \theta_I \le \theta \le \pi \end{cases}$$

The shapes of plastic zone for mode I crack under plane stress and plane strain cases are plotted in Fig. S5.13. It can be seen that the shapes of plastic zone for both cases are symmetric about the x-axis.

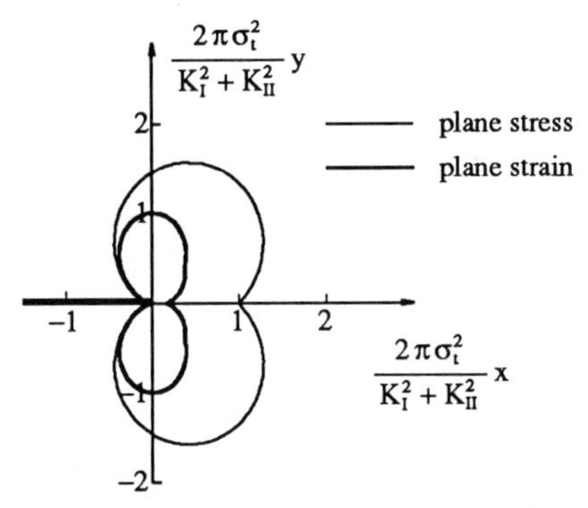

Figure S5.13. Shapes of plastic zone for Mode I using Tresca criterion

Prob. 5.14 Derive the expressions for the shapes of plastic zone near mode II crack and plot the shapes in the (x, y) plane. Use the Tresca criterion for both plane stress and plane strain cases.

Answer:

Plane Stress:

$$R(\theta) = \begin{cases} [-\sin\dfrac{\theta}{2} + \dfrac{1}{2}\sqrt{4 - 3\sin^2\theta}\,]^2 & -\pi \le \theta \le -\theta_{II} \\ 4 - 3\sin^2\theta & -\theta_{II} \le \theta \le \theta_{II} \\ [\ \sin\dfrac{\theta}{2} + \dfrac{1}{2}\sqrt{4 - 3\sin^2\theta}\,]^2 & \theta_{II} \le \theta \le \pi \end{cases}$$

Plane Strain

$$R(\theta) = 4 - 3\sin^2\theta$$

The shapes of plastic zone for mode II crack for plane stress and plane strain cases are plotted in Fig. S5.14.

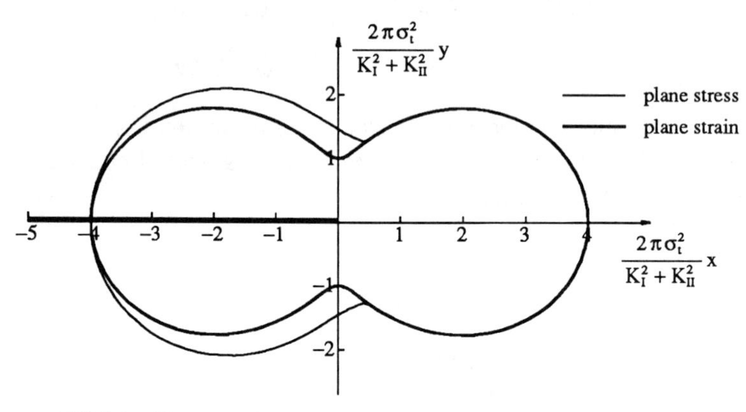

Figure S5.14. Shapes of plastic zone for Mode II using Tresca criterion

Prob. 5.15 Derive the expressions for the shapes of plastic zone near mode I crack and plot the shapes in the (x, y) plane. Use the von Mises criterion for both plane stress and plane strain cases.

Answer: The von Mises criterion can be written as

$$J_2^{1/2} - \frac{\sigma_t}{\sqrt{3}} = 0$$

Thus, using Equations (5.31) and (5.35), we express the radius of the plastic zone as

$$R(\theta) = 3\,\bar{J}_2$$

Substituting the stress expressions given in Probs. 5.9 and 5.10, the shape of the plastic zone for mode I is found to be

$$R(\theta) = \begin{cases} (1 + 3\sin^2\frac{\theta}{2})\cos^2\frac{\theta}{2} & \text{for plane stress} \\ [(1 - 2\nu)^2 + 3\sin^2\frac{\theta}{2}]\cos^2\frac{\theta}{2} & \text{for plane strain} \end{cases}$$

The shapes of plastic zone for mode I crack for both plane stress and plane strain cases are plotted in Fig. S5.15.

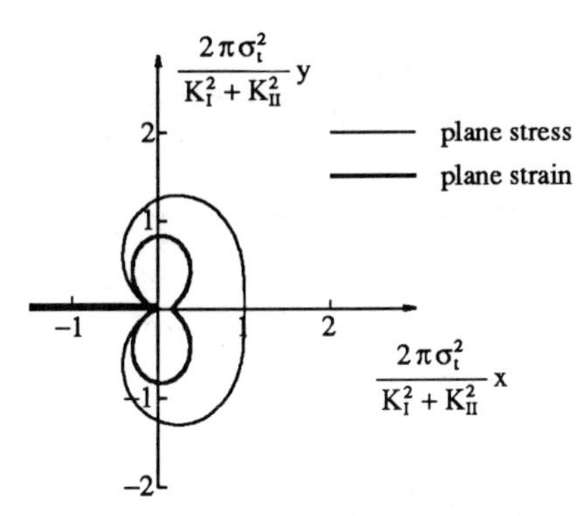

Figure S5.15. Shapes of plastic zone for Mode I using von Mises criterion

Prob. 5.16 Derive the expressions for the shapes of plastic zone near mode II crack and plot the shapes in the (x, y) plane. Use the von Mises criterion for both plane stress and plane strain cases.

Answer:

$$R(\theta) = \begin{cases} 3 - 8\sin^2\frac{\theta}{2} + 9\sin^4\frac{\theta}{2} & \text{for plane stress} \\ 3 - (8 + 4\nu - 4\nu^2)\sin^2\frac{\theta}{2} + 9\sin^4\frac{\theta}{2} & \text{for plane strain} \end{cases}$$

The shapes of plastic zone for mode II crack for both plane stress and plane strain cases are plotted in Fig. S5.16.

Prob. 5.17 Derive the expressions for the shapes of plastic zone near mode I crack and plot the shapes in the (x, y) plane. Use the Rankine criterion for both plane stress and plane strain cases.

Solution: The Rankine yield criterion has the form

$$\sigma_{max} = \sigma_t$$

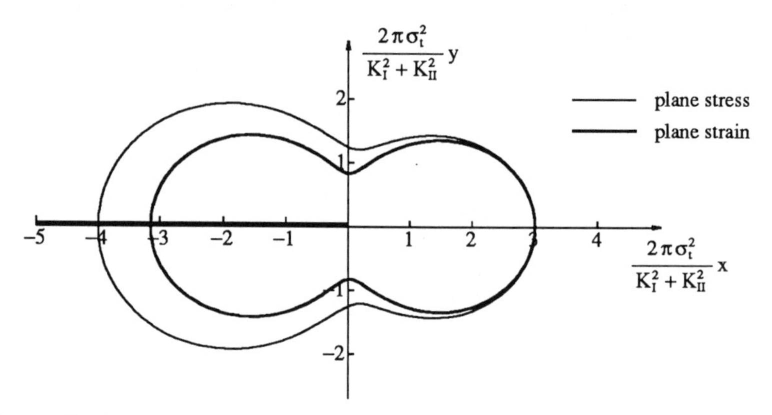

Figure S5.16. Shapes of plastic zone for Mode II using von Mises criterion

where σ_{max} is the maximum principal stress, and $\sigma_{max} \geq 0$. Using Equations (5.31) and (5.35), we obtain

$$R(\theta) = \bar{\sigma}_{max}^{-2} \tag{7.3}$$

For this criterion, there is no difference between plane strain and plane stress cases. Therefore, using the stress expressions given in Probs. 5.9 and 5.10, we obtain the expression for the plastic zone of mode I cracks as

$$R(\theta) = \begin{cases} \cos^2\frac{\theta}{2}\,(1-\sin\frac{\theta}{2})^2 & -\pi \leq \theta \leq 0 \\ \cos^2\frac{\theta}{2}\,(1+\sin\frac{\theta}{2})^2 & 0 \leq \theta \leq \pi \end{cases}$$

The shape of plastic zone for mode I crack is plotted in Fig. S5.17.

Prob. 5.18 Derive the expressions for the shapes of plastic zone near mode II crack and plot the shapes in the (x, y) plane. Use the Rankine criterion for both plane stress and plane strain cases.

Answer:

$$R(\theta) = \begin{cases} [-\sin\frac{\theta}{2} + \frac{1}{2}\sqrt{4-3\sin^2\theta}\;]^2 & -\pi \leq \theta \leq \theta_{II} \\ 0 & \theta_{II} \leq \theta \leq \pi \end{cases}$$

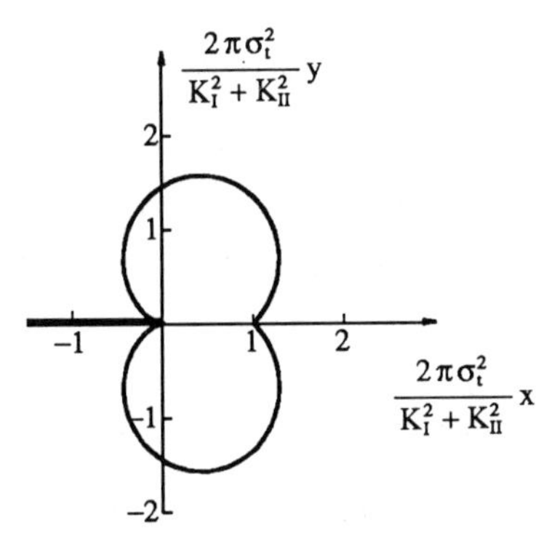

Figure S5.17. Shape of plastic zone of Mode I using Rankine criterion

The shape of plastic zone for mode I crack is plotted in Fig. S5.18.

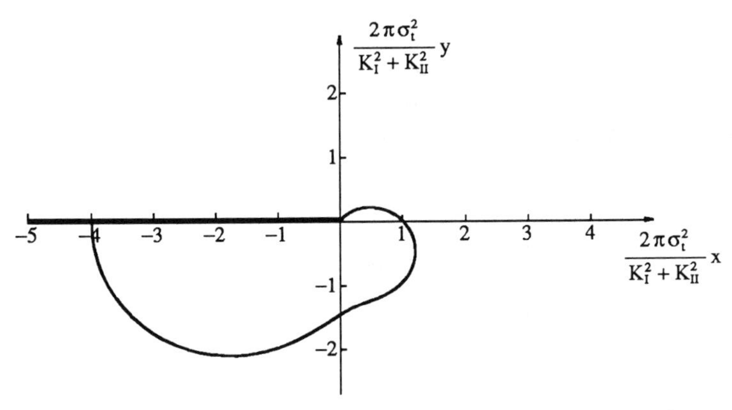

Figure S5.18. Shape of plastic zone for Mode II using Rankine criterion

Prob. 5.19 Derive the expressions for the shapes of plastic zone near mode I crack and plot the shapes in the (x, y) plane. Use the Mohr-Coulomb criterion for both plane stress and plane strain cases.

Solution: The Mohr-Coulomb criterion can be generally expressed as

$$m\,\sigma_{max} - \sigma_{min} = m\,\sigma_t$$

where σ_{max} and σ_{min} are the maximum and minimum principal stresses respectively, and m is ratio of compressive yield stress to tensile yield stress, $m = \sigma_c / \sigma_t$. Using Equations (5.31) and (5.35), we obtain

$$R(\theta) = \left[\frac{m\bar{\sigma}_{max} - \bar{\sigma}_{min}}{m} \right]^2$$

Using the stress expressions given in Probs. 5.9 and 5.10, the shapes of plastic zone for mode I crack are found to be

Plane Stress:

$$R(\theta) = \begin{cases} \cos^2\frac{\theta}{2}\,(1 - \sin\frac{\theta}{2})^2 & -\pi \le \theta \le 0 \\ \cos^2\frac{\theta}{2}\,(1 + \sin\frac{\theta}{2})^2 & 0 \le \theta \le \pi \end{cases}$$

Plane Strain:

$$R(\theta) = \begin{cases} \frac{1}{m^2}\cos^2\frac{\theta}{2}\,[(m-1) - (m+1)\sin\frac{\theta}{2}]^2 & -\pi \le \theta \le -\theta_I \\ \frac{1}{m^2}\cos^2\frac{\theta}{2}\,[m(1 - \sin\frac{\theta}{2}) - 2\nu]^2 & -\theta_I \le \theta \le 0 \\ \frac{1}{m^2}\cos^2\frac{\theta}{2}\,[m(1 + \sin\frac{\theta}{2}) - 2\nu]^2 & 0 \le \theta \le \theta_I \\ \frac{1}{m^2}\cos^2\frac{\theta}{2}\,[(m-1) + (m+1)\sin\frac{\theta}{2}]^2 & \theta_I \le \theta \le \pi \end{cases}$$

The shapes of plastic zone for mode I crack for both cases of plane stress and plane strain are plotted in Fig. S5.19 with $m = 2.5$.

Prob. 5.20 Derive the expressions for the shapes of plastic zone near mode II crack and plot the shapes in the (x, y) plane. Use the Mohr-Coulomb criterion for both plane stress and plane strain cases.

Answer:

Plane Stress:

$$R(\theta) = \begin{cases} [-\sin\frac{\theta}{2} + \frac{1}{2}\sqrt{4 - 3\sin^2\theta}\,]^2 & -\pi \le \theta \le -\theta_{II} \\ \frac{1}{m^2}[-(m-1)\sin\frac{\theta}{2} + \frac{(m+1)}{2}\sqrt{4 - 3\sin^2\theta}\,]^2 & -\theta_{II} \le \theta \le \theta_{II} \\ \frac{1}{m^2}[\ \sin\frac{\theta}{2} + \frac{1}{2}\sqrt{4 - 3\sin^2\theta}\,]^2 & \theta_{II} \le \theta \le \pi \end{cases}$$

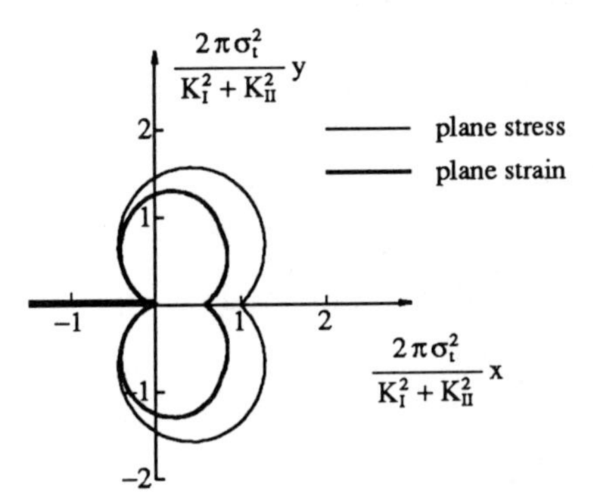

Figure S5.19. Shapes of plastic zone of Mode I using Mohr-Coulomb criterion

Plane Strain:

$$R(\theta) = \frac{1}{m^2}\left[-(m-1)\sin\frac{\theta}{2} + \frac{(m+1)}{2}\sqrt{4-3\sin^2\theta}\;\right]^2$$

The shapes of plastic zone for mode II crack for the cases of plane stress and plane strain are plotted in Fig. S5.20 with $m = 2.5$.

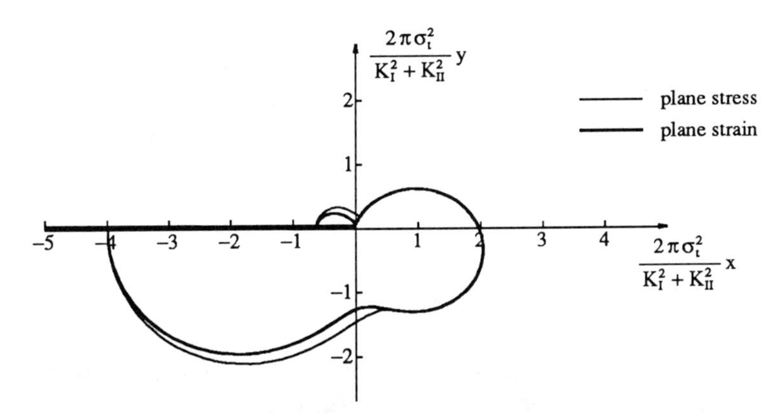

Figure S5.20. Shapes of plastic zone of Mode II using Mohr-Coulomb criterion

Prob. 5.21 Derive the expressions for the shapes of plastic zone near mode I crack and plot the shapes in the (x, y) plane. Use the Drucker-

Prager criterion for both plane stress and plane strain cases.

Solution: The Drucker-Prager criterion can be written as

$$\alpha I_1 + J_2^{1/2} = (\alpha + \frac{1}{\sqrt{3}})\,\sigma_t \tag{9.1}$$

Using Equations (5.31) and (5.35), we obtain

$$R(\theta) = \left[\frac{\alpha \bar{I}_1 + \bar{J}_2^{1/2}}{\alpha + \frac{1}{\sqrt{3}}} \right]^2$$

Using the stress expressions given in Probs. 5.9 and 5.10, the plastic zone for mode I can be expressed as

Plane Stress:

$$R(\theta) = \left[\frac{2\alpha \cos\frac{\theta}{2} + \frac{1}{\sqrt{3}} \cos\frac{\theta}{2}\,(1 + 3\sin^2\frac{\theta}{2})^{1/2}}{\alpha + \frac{1}{\sqrt{3}}} \right]^2$$

Plane Strain:

$$R(\theta) = \left[\frac{2\alpha\,(1+\nu) \cos\frac{\theta}{2} + \frac{1}{\sqrt{3}} \cos\frac{\theta}{2}\,[(1-2\nu)^2 + 3\sin^2\frac{\theta}{2}]^{1/2}}{\alpha + \frac{1}{\sqrt{3}}} \right]^2$$

The shapes of plastic zone for mode I crack for the cases of plane stress and plane strain are plotted in Fig. S5.21 with $\alpha = 0.2574$.

Prob. 5.22 Derive the expressions for the shapes of plastic zone near mode II crack and plot the shapes in the (x, y) plane. Use the Drucker-Prager yield criterion for both plane stress and plane strain cases.

Answer:
Plane Stress:

$$R(\theta) = \left[\frac{-2\alpha \sin\frac{\theta}{2} + \frac{1}{\sqrt{3}}\,(3 - 8\sin^2\frac{\theta}{2} + 9\sin^4\frac{\theta}{2})^{1/2}}{\alpha + \frac{1}{\sqrt{3}}} \right]^2$$

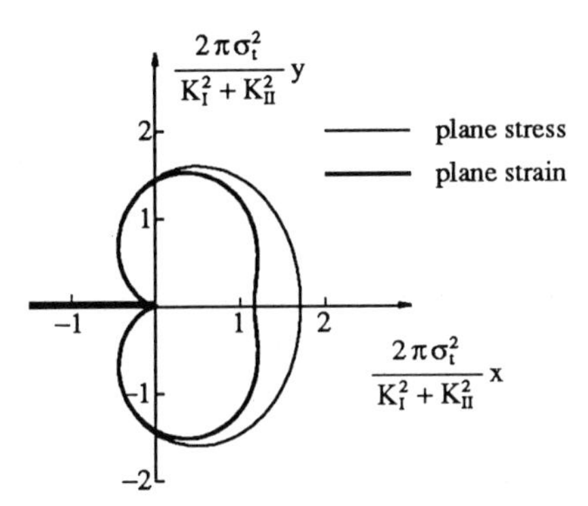

Figure S5.21. Shapes of plastic zone of Mode I using Drucker-Prager criterion

Plane Strain:

$$R(\theta) = \left[\frac{-2\alpha\,(1+\nu)\sin\frac{\theta}{2} + \frac{1}{\sqrt{3}}\,[3 - (8+4\nu-4\nu^2)\sin^2\frac{\theta}{2} + 9\sin^4\frac{\theta}{2}]^{1/2}}{\alpha + \frac{1}{\sqrt{3}}} \right]^2$$

The shapes of plastic zone for mode II crack for the cases of plane stress and plane strain are plotted in Fig. S5.22 with $\alpha = 0.2574$.

5.7 Thick-walled Vessel Problems

Prob. 5.23 A long thick-walled tube with open ends ($\sigma_z = 0$) is subjected to an internal pressure p shown in Fig. P5.23. The inner radius and outer radius are a and b respectively. Assuming the material follows the Tresca criterion with the uniaxial tensile strength σ_0.

a. Determine the elastic limit internal pressure p_e;
b. Determine the relationship between the elastic-plastic boundary and the internal pressure p for $p > p_e$;
c. Determine the plastic limit internal pressure p_c; and
d. For the case of $b/a = 2$, plot σ_r and σ_θ vs. r curves for the elastic-plastic boundary at $c = a$, $c = (a+b)/2$, and $c = b$ respectively.

Solution: The equilibrium and compatibility equations for this problem are, in the usual notation

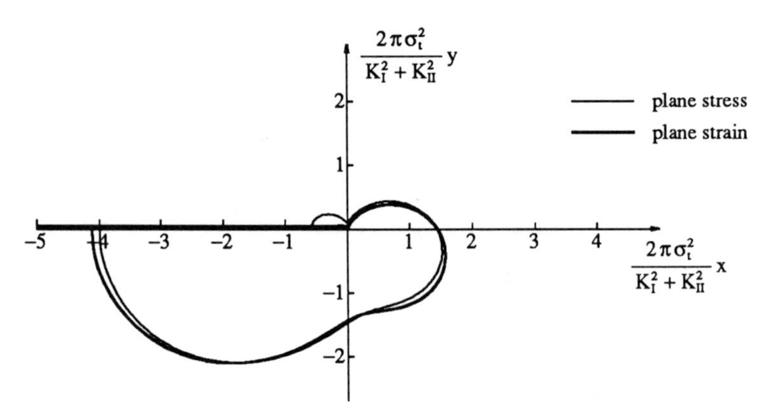

Figure S5.22. Shapes of plastic zone of Mode II using Drucker-Prager criterion

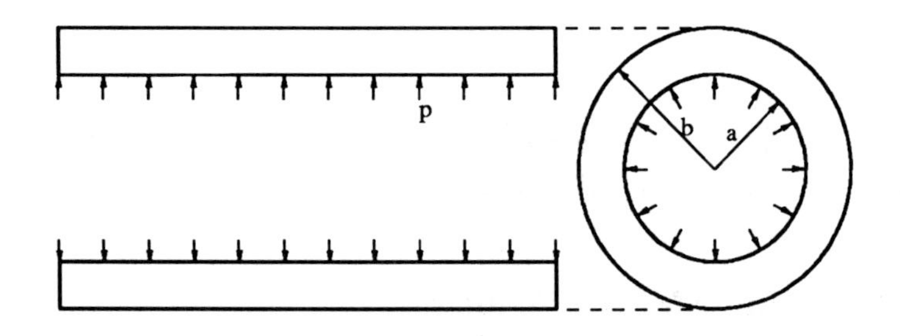

Figure P5.23.

$$\frac{d\sigma_r}{dr} - \frac{\sigma_\theta - \sigma_r}{r} = 0, \quad \sigma_z = 0, \quad \tau_{\theta z} = \tau_{rz} = \tau_{r\theta} = 0 \tag{1}$$

$$\varepsilon_r = \frac{du}{dr}, \quad \varepsilon_\theta = \frac{u}{r}, \quad \gamma_{\theta z} = \gamma_{rz} = \gamma_{r\theta} = 0 \tag{2}$$

The boundary conditions arc

$$\sigma_r \,|_{r=a} = -p, \quad \sigma_r \,|_{r=b} = 0 \tag{3}$$

Combining Equations (1) and (2) with the elastic stress-strain relation, we obtain the basic solution for the thick-walled tube problem

$$\sigma_r = \frac{A}{r^2} + B, \quad \sigma_\theta = -\frac{A}{r^2} + B \tag{4}$$

Utilizing the boundary conditions (3), the elastic solution is obtained as

$$\sigma_r = \bar{p}\,(1 - \frac{b^2}{r^2}) < 0, \quad \sigma_\theta = \bar{p}\,(1 + \frac{b^2}{r^2}) > 0, \quad \sigma_z = 0 \tag{5}$$

where

$$\bar{p} = \frac{a^2 p}{b^2 - a^2}$$

(a) The Tresca criterion for this case is

$$\sigma_\theta - \sigma_r = \sigma_0 \tag{6}$$

Substituting the elastic solution into Eq. (6), and noting that the left-hand-side of this equation has its maximum value at $r = a$, we find that the tube begins to yield at inner face, and has the elastic limit pressure

$$p_e = \frac{\sigma_0}{2}\,(1 - \frac{a^2}{b^2})$$

(b) For the case $p > p_e$, denote the elastic-plastic boundary by $r = c$. Thus, for $a \le r \le c$, the tube is in a plastic state. By utilizing the yield condition (5), Equilibrium equation (1) becomes

$$\frac{d\sigma_r}{dr} - \frac{\sigma_0}{r} = 0 \tag{7}$$

Solving the equation and using the boundary condition at $r = a$, we obtain the stress distribution in the plastic zone as

$$\sigma_r^{(1)} = \sigma_0\, ln\,\frac{r}{a} - p, \quad \sigma_\theta^{(1)} = \sigma_0\,(1 + ln\,\frac{r}{a}) - p \tag{8}$$

where the sup-script (1) denotes the stress components in the plastic zone.

In the elastic zone, $c \le r \le b$, the basic elastic solution (4) remains valid. By utilizing the boundary condition at $r = b$, we express the stress distribution in the elastic zone as

$$\sigma_r^{(2)} = A\,(\frac{1}{r^2} - \frac{1}{b^2}), \quad \sigma_\theta^{(2)} = -A\,(\frac{1}{r^2} + \frac{1}{b^2}) \tag{9}$$

where the sup-script (2) denotes the stress components in the elastic zone. Utilizing the continuity condition at $r = c$,

$$\sigma_r^{(1)}\,|_{r=c} = \sigma_r^{(2)}\,|_{r=c}, \qquad (\sigma_\theta^{(2)} - \sigma_r^{(2)})\,|_{r=c} = \sigma_0 \tag{10}$$

the constant A is determined and the pressure p vs. c relationship becomes

$$p = \sigma_0 \, ln \frac{c}{a} + \frac{\sigma_0}{2}(1 - \frac{c^2}{b^2}) \tag{11}$$

The complete stress distribution for $p \geq p_e$ is

$$\sigma_r^{(1)} = \sigma_0 \, ln \frac{r}{a} - p, \quad \sigma_\theta^{(1)} = \sigma_0 (1 + ln \frac{r}{a}) - p, \qquad a \leq r \leq c$$

$$\sigma_r = -\frac{c^2 \sigma_0}{2}(\frac{1}{r^2} - \frac{1}{b^2}), \quad \sigma_\theta = \frac{c^2 \sigma_0}{2}(\frac{1}{r^2} + \frac{1}{b^2}), \quad c \leq r \leq b$$

(c) Let $c \to b$ in Equation (11), we obtain the plastic limit load as

$$p_c = \sigma_0 \, ln \frac{b}{a}$$

(d) The stress vs. r curves for several values of c are plotted in Fig. S5.23.

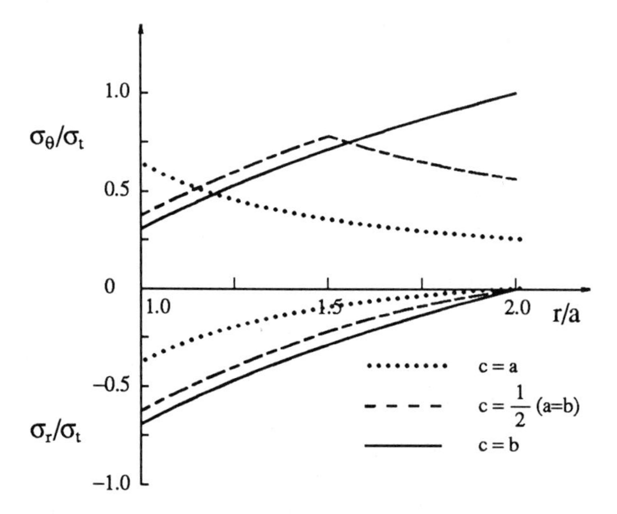

Figure S5.23.

Prob. 5.24 Same thick-walled tube problem as in Prob. 5.23. Assume the tube is made of concrete materials and use the Rankine criterion with uniaxial tensile strength σ_t.

Solution:

(a) The general elastic solution for the thick-walled tube has been obtained in Prob. 5.23 as

$$\sigma_r = \frac{A}{r^2} + B \tag{1}$$

$$\sigma_\theta = -\frac{A}{r^2} + B \tag{2}$$

Using the boundary conditions

$$\sigma_r \,|_{r=a} = -p \tag{3}$$

$$\sigma_r \,|_{r=b} = 0, \tag{4}$$

we obtain the elastic solution

$$\sigma_r = \bar{p}\,(1 - \frac{b^2}{r^2}) < 0 \tag{5}$$

$$\sigma_\theta = \bar{p}\,(1 + \frac{b^2}{r^2}) > 0 \tag{6}$$

$$\sigma_z = 0$$

where

$$\bar{p} = \frac{a^2 p}{b^2 - a^2}$$

The Rankine criterion for this case has the simple form

$$\sigma_\theta = \sigma_t \tag{7}$$

Substituting this into Eq. (6), we obtain the elastic limit pressure,

$$p_e = \frac{b^2 - a^2}{b^2 + a^2}\,\sigma_t \tag{8}$$

(b) For the case $p > p_e$, we denote the elastic-plastic boundary by $r = c$. Then, for $a \le r \le c$, the equilibrium equation has the form

$$\frac{d\sigma_r^{(1)}}{dr} + \frac{\sigma_r^{(1)}}{r} - \frac{\sigma_t}{r} = 0 \tag{9}$$

where the sup-script (1) denotes the stress components in the plastic region. Solving Equation (9), we obtain

$$\sigma_r^{(1)} = \frac{D}{r} + \sigma_t$$

where D is an integration constant. Using the boundary condition (3), the constant is found to be

$$D = -a\,(p + \sigma_t)$$

For $a \le r \le c$, we have

$$\sigma_r^{(1)} = \frac{D}{r} + \sigma_t = -\frac{a}{r}\,(p + \sigma_t) + \sigma_t \tag{10a}$$

$$\sigma_\theta^{(1)} = \sigma_t \tag{10b}$$

For $c \le r \le b$, the stress state remains in an elastic state. Using the general solutions, Eqs. (1) and (2), and the boundary condition, Eq. (4), we have

$$\sigma_r^{(2)} = A\left(\frac{1}{r^2} - \frac{1}{b^2}\right) \tag{11}$$

$$\sigma_\theta^{(2)} = -A\left(\frac{1}{r^2} + \frac{1}{b^2}\right) \tag{12}$$

where the sup-script (2) denotes the stress components in the elastic region, and A is a constant.

Using the continuity conditions at $r = c$,

$$\sigma_r^{(1)}\big|_{r=c} = \sigma_r^{(2)}\big|_{r=c}, \quad \sigma_\theta^{(2)}\big|_{r=c} = \sigma_t$$

we determine the constant

$$A = \frac{\sigma_t\left(1 - \frac{a}{c}\right) - \frac{a}{c}\,p}{\frac{1}{c^2} - \frac{1}{b^2}} = \frac{-c^2 b^2}{b^2 + c^2}\,\sigma_t \tag{13}$$

and lead to the p vs. c relation as

$$\frac{a}{c}\,p = \sigma_t\left(1 - \frac{a}{c}\right) + \frac{b^2 - c^2}{b^2 + c^2}\,\sigma_t \tag{14}$$

The relationship (14) can also be obtained by the conditions

$$\sigma_r^{(1)}\big|_{r=c} = \sigma_r^{(2)}\big|_{r=c}, \quad \text{and} \quad \sigma_r^{(2)}\big|_{r=c} = -q \tag{15}$$

where q is the elastic limit pressure as given by Eq. (8) with a being replaced by c,

$$q = \frac{b^2 - c^2}{b^2 + c^2}\,\sigma_t \tag{16}$$

(c) Letting the elastic-plastic boundary at $r = b$ and replacing c in Equation (14) by b, we obtain

$$\frac{a}{b}\,p_c = \sigma_t\left(1 - \frac{a}{b}\right)$$

or

$$p_c = \sigma_t\left(\frac{b}{a} - 1\right) \tag{17}$$

(d) To plot the σ_r and σ_θ vs. r curves, we first determine the stress expressions for three cases. For the given ratio $b/a = 2$, we have

$$p_e = \frac{3}{5}\,\sigma_t, \quad p_c = \sigma_t$$

For the case c = a, the tube is in an elastic state except at the radius r = a. We therefore use the elastic solution, Eqs. (5) and (6),

$$\sigma_r = \bar{p}\,(1 - \frac{b^2}{r^2}), \qquad \sigma_\theta = \bar{p}\,(1 + \frac{b^2}{r^2})$$

and

$$\bar{p} = \frac{a^2 p}{b^2 - a^2} = \frac{1}{3}\,p_e = \frac{1}{5}\,\sigma_t$$

For the case $c = \frac{1}{2}(a+b)$, in the plastic region, $a \le r \le \frac{1}{2}(a+b)$, we use Eqs. (10a) and (10b) and obtain

$$\sigma_r = \sigma_t - \frac{a}{r}\,(p + \sigma_t), \qquad \sigma_\theta = \sigma_t$$

and in the elastic region, $\frac{1}{2}(a+b) \le r \le b$, we use Eqs. (11) and (12) and obtain

$$\sigma_r = A\,(\frac{1}{r^2} - \frac{1}{b^2}), \qquad \sigma_\theta = -A\,(\frac{1}{r^2} + \frac{1}{b^2})$$

where the constant A and the internal pressure p can be determined by Eqs. (13) and (14)

$$A = -\frac{36}{25}\,a^2\sigma_t, \qquad p = \frac{46}{50}\,\sigma_t$$

For the case c = b, the tube is in a plastic state, and the internal pressure has the value $p = p_c = \sigma_t$. Using Eqs. (10a) and (10b), we have

$$\sigma_r = \sigma_t - \frac{a}{r}\,(\sigma_t + \sigma_t) = (1 - \frac{2a}{r})\,\sigma_t = (1 - \frac{b}{r})\,\sigma_t$$

$$\sigma_\theta = \sigma_t$$

All the curves for these three cases are plotted in Figure S5.24.

Prob. 5.25 Same thick-walled problem as in Prob. 5.24. Use the Mohr-Coulomb criterion with the uniaxial tensile and compression strengths being σ_t and σ_c respectively, and $\sigma_c / \sigma_t = 10$.

Solution:

(a) Using the general elastic solution for a thick-walled tube,

$$\sigma_r = \frac{A}{r^2} + B \tag{1}$$

$$\sigma_\theta = -\frac{A}{r^2} + B \tag{2}$$

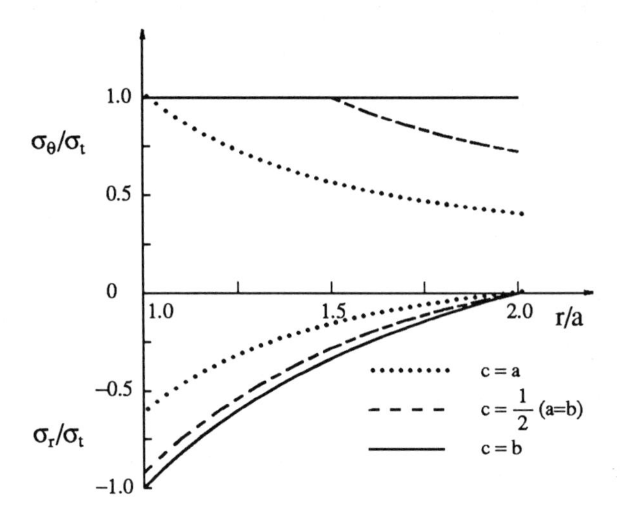

Figure S5.24.

and the boundary conditions

$$\sigma_r \,|_{r=a} = -p \tag{3}$$

$$\sigma_r \,|_{r=b} = 0 \tag{4}$$

we obtain the elastic solution

$$\sigma_r = \bar{p}\,(1 - \frac{b^2}{r^2}) < 0 \tag{5}$$

$$\sigma_\theta = \bar{p}\,(1 + \frac{b^2}{r^2}) > 0 \tag{6}$$

$$\sigma_z = 0$$

where

$$\bar{p} = \frac{a^2 p}{b^2 - a^2}$$

The Mohr-Coulomb criterion for this case is

$$m\,\sigma_\theta - \sigma_r = \sigma_c \tag{7}$$

where $m = \sigma_c / \sigma_t$. Substituting Equations (5) and (6) to this equation, and letting $r = a$, we obtain the elastic limit pressure,

$$p_e = \frac{(b^2 - a^2)}{m(a^2 + b^2) + (b^2 - a^2)}\,\sigma_c \tag{8}$$

(b) For the case $p > p_e$, denote the elastic-plastic boundary by $r = c$. Then, for $a \le r \le c$, the equilibrium equation has the form

$$m \frac{d\sigma_r^{(1)}}{dr} + (m-1)\frac{\sigma_r^{(1)}}{r} - \frac{\sigma_c}{r} = 0. \tag{9}$$

where, again, the sup-script (1) denotes the stress components in the plastic region. Solving Eq. (9), we obtain

$$\sigma_r^{(1)} = D\, r^{-n} + \frac{\sigma_t}{n} = D\, r^{-n} + \frac{\sigma_c}{nm}$$

where $n = \dfrac{m-1}{m}$, and D is an integration constant. Using the boundary condition (3), we find

$$D = -\left(p + \frac{\sigma_t}{n}\right) a^n$$

Therefore, for $a \le r \le c$, we have

$$\sigma_r^{(1)} = -\left(p + \frac{\sigma_t}{n}\right)\left(\frac{a}{r}\right)^n + \frac{\sigma_t}{n} \tag{10a}$$

$$\sigma_\theta^{(1)} = -\frac{1}{m}\left(p + \frac{\sigma_t}{n}\right)\left(\frac{a}{r}\right)^n + \frac{\sigma_t}{mn} + \sigma_t \tag{10b}$$

For $r \ge c$, the stress state remains in an elastic state. Using the general solution, Eqs. (1) and (2), and the boundary conditions (4), we obtain

$$\sigma_r^{(2)} = A\left(\frac{1}{r^2} - \frac{1}{b^2}\right) \tag{11}$$

$$\sigma_\theta^{(2)} = -A\left(\frac{1}{r^2} + \frac{1}{b^2}\right) \tag{12}$$

where the sup-script (2) denotes the stress components in the elastic region, and A is a constant.

Using the continuity conditions at $r = c$,

$$\sigma_r^{(1)}\big|_{r=c} = \sigma_r^{(2)}\big|_{r=c}, \qquad (m\,\sigma_\theta^{(2)} - \sigma_r^{(2)})\big|_{r=c} = \sigma_c$$

we determine the constant

$$A = -\frac{\sigma_c\, c^2 b^2}{m(b^2 + c^2) + (b^2 - c^2)} = -\frac{\left[\left(p + \frac{\sigma_t}{n}\right)\left(\frac{a}{c}\right)^n - \frac{\sigma_t}{n}\right] c^2 b^2}{b^2 - c^2}, \tag{13}$$

and lead to the p vs. c relation

$$\left(\frac{a}{c}\right)^n p = \frac{\sigma_t}{n}\left[1 - \left(\frac{a}{c}\right)^n\right] + \frac{m\,\sigma_t\,(b^2 - c^2)}{m(b^2 + c^2) + (b^2 - c^2)} \tag{14}$$

(c) Letting the elastic-plastic boundary at $r = b$, and replacing c in Eq. (14) by b, we obtain the plastic limit pressure as

$$(\frac{a}{b})^n \, p_c = \frac{\sigma_t}{n} \, [1 - (\frac{a}{b})^n]$$

or

$$p_c = \frac{\sigma_t}{n} \, [(\frac{b}{a})^n - 1] \tag{15}$$

(d) For the given ratios $b/a = 2$, $m = \sigma_c/\sigma_t = 10$, we have $n = \frac{m-1}{m} = \frac{9}{10}$ and find

$$p_e = \frac{3}{53} \, \sigma_c = \frac{30}{53} \, f_t,$$

$$p_c = \frac{10\sigma_t}{9} \, [2^{9/10} - 1] = \frac{\sigma_c}{9} \, [2^{9/10} - 1]$$

For the case $c = a$, the tube is in an elastic stress state except at the radius $r = a$. We therefore use the elastic solution, Eqs. (5) and (6),

$$\sigma_r = \bar{p}\,(1 - \frac{b^2}{r^2}), \qquad \sigma_\theta = \bar{p}\,(1 + \frac{b^2}{r^2})$$

and find

$$\bar{p} = \frac{a^2}{b^2 - a^2} \, p_e = \frac{1}{3} \, p_e = \frac{10}{53} \, \sigma_t$$

For the case $c = \frac{1}{2}(a+b)$, in the plastic region, $a \le r \le \frac{1}{2}(a+b)$, we use Equations (10a) and (10b), and obtain

$$\sigma_r = \frac{10}{9} \, \sigma_t - (p + \frac{10}{9} \, \sigma_t) \, (\frac{a}{r})^{9/10}$$

$$\sigma_\theta = \frac{10}{9} \, \sigma_t - (\frac{1}{10} \, p + \frac{1}{9} \, \sigma_t) \, (\frac{a}{r})^{9/10}$$

and in the elastic region, $\frac{1}{2}(a + b) \le r \le b$, we use Eqs. (11) and (12), and obtain

$$\sigma_r = A\,(\frac{1}{r^2} - \frac{1}{b^2}), \qquad \sigma_\theta = -A\,(\frac{1}{r^2} + \frac{1}{b^2})$$

where the constant A and internal pressure p are determined by Eqs. (13) and (14)

$$A = -\frac{360}{257} \, \sigma_t$$

$$p = \frac{10}{9}\left[\left(\frac{3}{2}\right)^{9/10} - 1\right]\sigma_t + \left(\frac{3}{2}\right)^{9/10}\frac{70}{257}\sigma_t$$

For the case $c = b$, the tube is in a plastic state, and the internal pressure has the value $p = p_s$. Using Eqs. (10a) and (10b), we have

$$\sigma_r = \frac{10}{9}\sigma_t\left[1 - \left(\frac{2a}{r}\right)^{9/10}\right] = \frac{10}{9}\sigma_t\left[1 - \left(\frac{b}{r}\right)^{9/10}\right]$$

$$\sigma_\theta = \sigma_t + \frac{1}{9}\sigma_t\left[1 - \left(\frac{2a}{r}\right)^{9/10}\right] = \frac{1}{9}\sigma_t\left[10 - \left(\frac{b}{r}\right)^{9/10}\right]$$

All the curves for these three cases are plotted in Figure S5.25.

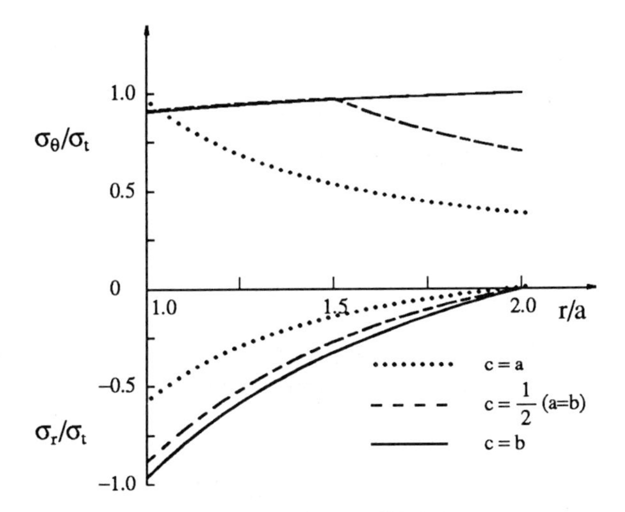

Figure S5.25.

Prob. 5.26 A long vertical circular hole in a half space of rock with an inner radius a is subjected to an internal pressure p as shown in Fig. P5.26. Using the Tresca criterion, with the unaxial tensile strength σ_0, determine the relationship between the radius of the plastic zone and the internal pressure.

Answer: $p = \frac{\sigma_0}{2} + \sigma_0 \, ln \frac{c}{a}$

Prob. 5.27 Same problem as in Prob. 5.26. Use the Rankine criterion with the uniaxial tensile strength being σ_t.

Figure P5.26.

Answer: $p = \sigma_t \left(2\frac{c}{a} - 1\right)$

Prob. 5.28 Same problem as in Prob. 5.26. Use the Mohr-Coulomb criterion with the uniaxial tensile and compression strengths being σ_t and σ_c.

Answer: $p = \frac{m}{m+1}\sigma_t + \frac{\sigma_t}{n}\left[1 - \left(\frac{a}{c}\right)^n\right]$, where $n = \frac{m-1}{m}$

Prob. 5.29 A composite of n tubes made of the same material, one inside the other is shown in Fig. P5.29. The inner and outer radii of the n tubes are (r_i, r_1), (r_1, r_2), ..., (r_{n-1}, r_e), respectively. The composite tube is subjected to an internal pressure p. Use the Tresca yield criterion. Assuming the yielding occurs simultaneously at the inner surfaces of each tube, show that

a. The inner pressure for the initial yielding is given by

$$p_e = \frac{\sigma_0}{2}\left\{n - \left[\left(\frac{r_i}{r_1}\right)^2 + \left(\frac{r_1}{r_2}\right)^2 + \cdots + \left(\frac{r_{n-1}}{r_e}\right)^2\right]\right\}$$

in which σ_0 is the yield stress in simple tension.

b. If the ratio of the outer and inner radii of each tube is

$$\frac{r_k}{r_{k-1}} = \left(\frac{r_e}{r_i}\right)^{\frac{1}{n}} \quad (k = 0, 1, 2, \cdots n; \text{ and } r_0 = r_i,\ r_n = r_e)$$

the pressure p_e takes the maximum value for the initial yielding, and

$$(p_e)_{max} = \frac{n\sigma_0}{2}\left[1 - \left(\frac{r_i}{r_e}\right)^{2/n}\right]$$

c. The plastic limit pressure p_c is given by

$$p_c = \sigma_0 \, ln \, \frac{r_e}{r_i}$$

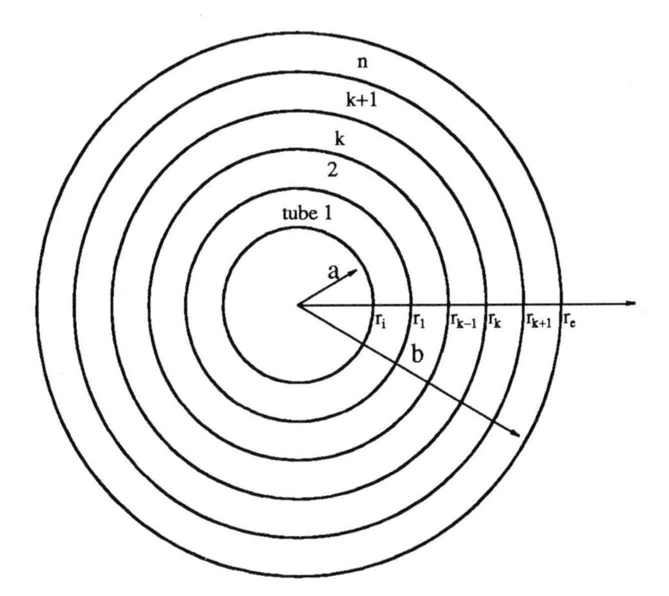

Figure P5.29.

Solution: The general elastic solution of the composite tube is

$$\sigma_r^{(k)} = A_k - \frac{1}{r^2} B_k, \qquad \sigma_\theta^{(k)} = A_k + \frac{1}{r^2} B_k \qquad (k = 1, 2, ... n) \tag{1}$$

The boundary and continuity conditions are

$$\sigma_r^{(1)} \,|_{r=r_i} = -p, \qquad \sigma_r^{(n)} \,|_{r=r_e} = 0 \tag{2}$$

$$\sigma_r^{(k)} \,|_{r=r_k} = \sigma_r^{(k+1)} \,|_{r=r_k}, \quad (k = 1,2,...n-1) \tag{3}$$

There are a total of (n+1) conditions.

(a) Assuming yielding occurs simultaneously at the inner surfaces of each tube, the following n conditions need be used

$$(\sigma_\theta^{(k)} - \sigma_r^{(k)})\big|_{r=r_{k-1}} = \sigma_0, \qquad (k = 1,2,...n) \tag{4}$$

Since we have a total of (2n+1) conditions, Equations. (2), (3) and (4), for (2n+1) unknowns. Thus, p_e, A_k, B_k (k = 1,2...n) can be determined uniquely. Substituting Equation (1) to Equation (4), we obtain

$$B_k = \frac{\sigma_0}{2} r_{k-1}^2, \quad (k = 1,2,...n) \tag{5}$$

Using the second boundary condition in Equation (2), we have

$$\sigma_r^{(n)}\big|_{r=r_e} = A_n - \frac{1}{r_e^2} B_n = A_n - \frac{\sigma_0^2}{2}\frac{r_{n-1}^2}{r_e^2} = 0, \tag{6}$$

and obtain

$$A_n = \frac{\sigma_0}{2}\frac{r_{n-1}^2}{r_e^2}$$

Using the continuity condition (3), at $r = r_{n-1}$, we have

$$\sigma_r^{(n-1)}\big|_{r=r_{n-1}} = \sigma_r^{(n)}\big|_{r=r_{n-1}}$$

or

$$A_{n-1} - \frac{1}{r_{n-1}^2} B_{n-1} = A_n - \frac{1}{r_{n-1}^2} B_n$$

we obtain

$$A_{n-1} = A_n + \frac{1}{r_{n-1}^2} B_{n-1} - \frac{1}{r_{n-1}^2} B_n = -\frac{\sigma_0}{2}\left[1 - \frac{r_{n-2}^2}{r_{n-1}^2} - \frac{r_{n-1}^2}{r_e^2}\right], \tag{7}$$

Now, we will use the mathematical induction and show that

$$A_k = -\frac{\sigma_0}{2}\left[(n-k) - \frac{r_{k-1}^2}{r_k^2} - \frac{r_k^2}{r_{k+1}^2} - \cdots - \frac{r_{n-1}^2}{r_e^2}\right] \tag{8}$$

First, Eq. (8) is valid for $k = n - 1$. Assume Eq. (8) is valid for k. Considering the continuity condition (3), at $r = r_{k-1}$, we have

$$\sigma_r^{(k-1)}\big|_{r=r_{k-1}} = \sigma_r^{(k)}\big|_{r=r_{k-1}}$$

Using Eqs. (1) and (5), we obtain

$$A_{k-1} = \frac{1}{r_{k-1}^2} B_{k-1} + A_k - \frac{1}{r_{k-1}^2} B_k = A_k + \frac{\sigma_0}{2}\frac{r_{k-2}^2}{r_{k-1}^2} - \frac{\sigma_0}{2}$$

$$= -\frac{\sigma_0}{2}\left[(n-k+1) - \frac{r_{k-2}^2}{r_{k-1}^2} - \frac{r_{k-1}^2}{r_k^2} - \cdots - \frac{r_{n-1}^2}{r_e^2}\right]$$

Therefore, Eq. (8) is proved.

Using the boundary condition at $r = r_i$, Eq. (2), we obtain

$$A_1 - \frac{1}{r_i^2} B_1 = -p_e, \quad \text{or} \quad p_e = -A_1 + \frac{\sigma_0}{2}$$

Substituting Eq. (8) for A_1 into the above equation, we obtain

$$p_e = \frac{\sigma_0}{2} [(n-1) - \frac{r_i^2}{r_1^2} - \frac{r_1^2}{r_2^2} - \cdots - \frac{r_k^2}{r_{k+1}^2} - \cdots - \frac{r_{n-1}^2}{r_e^2}] + \frac{\sigma_0}{2}$$

$$= \frac{\sigma_0}{2} \{n - [(\frac{r_i}{r_1})^2 + (\frac{r_1}{r_2})^2 + \cdots + (\frac{r_k}{r_{k+1}})^2 + \cdots + (\frac{r_{n-1}}{r_e})^2]\} \tag{9}$$

(b) To find the maximum value of the elastic limit pressure p_e, let $\partial p_e / \partial r_k = 0$, and we obtain

$$\frac{\sigma_0}{2} [2 \frac{r_{k-1}^2}{r_k^3} - 2 \frac{r_k}{r_{k+1}^2}] = 0$$

or

$$\frac{r_{k-1}^2}{r_k^3} = \frac{r_k}{r_{k+1}^2}, \quad \text{or} \quad \frac{r_{k-1}^2}{r_k^2} = \frac{r_k^2}{r_{k+1}^2}$$

Therefore, if $r_{k-1} / r_k = \text{constant}$, $(k = 1,2,...,n)$, p_e will take the maximum value. Let

$$\frac{r_{k-1}}{r_k} = \alpha = \text{constant}$$

we obtain

$$r_e = \alpha r_{n-1} = \alpha^2 r_{n-1} = \cdots = \alpha^{n-1} r_1 = \alpha^n r_i$$

or

$$\alpha = (\frac{r_e}{r_i})^{1/n} = \frac{r_{k-1}}{r_k}$$

Using Eq. (9), the maximum value of p_e is obtained as

$$(p_e)_{max} = \frac{\sigma_0}{2} [n - \frac{n}{\alpha^2}] = \frac{n\sigma_0}{2} [1 - \frac{1}{\alpha^2}] = \frac{n\sigma_0}{2} [1 - (\frac{r_i}{r_e})^{2/n}]$$

(c) The general elastic-plastic solution of each tube corresponding to the plastic collapse pressure p_c is

$$\sigma_r^{(k)} = \sigma_0 \, ln \, r + c_k, \quad \sigma_\theta^{(k)} = \sigma_0 + \sigma_r^{(k)}, \quad (k = 1,2,...,n) \tag{11}$$

There are (n + 1) unknowns, p_c, c_k $(k = 1,2,...n)$ and (n + 1) boundary and continuity conditions, Eqs. (2) and (3). Using the boundary conditions at $r = r_e$, we obtain

$$c_n = -\sigma_0 \, ln\, r_e$$

Using the continuity condition at $r = r_{n-1}$, we have

$$\sigma_0 \, ln\, r_{n-1} + c_{n-1} = \sigma_0 \, ln\, r_{n-1} + c_n$$

from which we obtain

$$c_{n-1} = c_n \tag{12}$$

It can be shown that the recurrence formula $c_{k-1} = c_k$ is true. Therefore, we have

$$c_1 = c_n = -\sigma_0 \, ln\, r_e$$

Using the boundary condition at $r = r_i$, we obtain

$$\sigma_0 \, ln\, r_i + c_1 = -p_c$$

or

$$p_c = -\sigma_0 \, ln\, r_i + \sigma_0 \, ln\, r_e = \sigma_0 \, ln\, \frac{r_e}{r_i} \tag{13}$$

Prob. 5.30 A thick-walled tube with open ends is subjected to an internal pressure and axial force $P = \alpha (b^2 - a^2) \pi p$. The tube yields at $r = a$ when $p = p_0$ without axial force ($\alpha = 0$). Determine the elastic limit load P_e in terms of α and p_0. Assume the Tresca criterion.

Answer:

$$p_y = \begin{cases} \dfrac{1 + \dfrac{b^2 + a^2}{b^2 - a^2}}{-\alpha + \dfrac{b^2 + a^2}{b^2 - a^2}} \, p_0 & \alpha < -1 \\[2ex] p_0 & -1 \le \alpha < \dfrac{b^2 + a^2}{b^2 - a^2} \\[2ex] \dfrac{1 + \dfrac{b^2 + a^2}{b^2 - a^2}}{1 + \alpha} \, p_0 & \dfrac{b^2 + a^2}{b^2 - a^2} \le \alpha \end{cases}$$

Prob. 5.31 Same thick-walled tube problem as in Prob. 5.30. Use the Mohr-Coulomb criterion with a known parameter m.

Answer:

$$p_y = \begin{cases} \dfrac{1 + m\dfrac{b^2+a^2}{b^2-a^2}}{-\alpha + m\dfrac{b^2+a^2}{b^2-a^2}}\, p_0 & \alpha < -1 \\[2ex] p_0 & -1 \le \alpha < \dfrac{b^2+a^2}{b^2-a^2} \\[2ex] \dfrac{1 + m\dfrac{b^2+a^2}{b^2-a^2}}{1 + m\alpha}\, p_0 & \dfrac{b^2+a^2}{b^2-a^2} \le \alpha \end{cases}$$

Prob. 5.32 A thick-walled tube with open ends is subjected to an external pressure p. Determine the elastic limit load P_e and the plastic collapse load P_c assuming the Tresca criterion.

Answer: $p_e = \dfrac{\sigma_0}{2}(1 - \dfrac{a^2}{b^2}), \quad p_c = \sigma_0 (1 - \dfrac{a}{b})$.

Prob. 5.33 A thick-walled sphere with an inner radius a and an outer radius b is subjected to an internal pressure p. Using the Tresca criterion,

a. Find the elastic limit pressure p_e; and
b. Find the elastic-plastic solution and the plastic limit pressure p_c.

Solution: The basic equations for problems with spherical symmetry are

Equilibrium: $$\frac{d\sigma_r}{dr} + \frac{2}{r}(\sigma_r - \sigma_\theta) = 0, \quad \sigma_\theta = \sigma_\phi, \ \tau_{r\theta} = \tau_{r\phi} = \tau_{\theta\phi} = 0 \tag{1}$$

Compatability: $$\varepsilon_r = \frac{du}{dr}, \ \varepsilon_\theta = \frac{u}{r}, \ \varepsilon_\phi = \varepsilon_\theta, \ \gamma_{r\theta} = \gamma_{r\phi} = \gamma_{\theta\phi} = 0 \tag{2}$$

Elastic stress-strain relation:

$$\sigma_r = (K + \frac{4}{3}G)\,\varepsilon_r + (2K - \frac{4}{3}G)\,\varepsilon_\theta$$

$$\sigma_\theta = (2K + \frac{2G}{3})\,\varepsilon_\theta + (K - \frac{2}{3}G)\,\varepsilon_r \tag{3}$$

Substituting Eqs. (2) and (3) into Eq. (1), we obtain the equilibrium equation expressed in terms of displacement u as

$$\frac{d^2u}{dr^2} + \frac{2}{r}\frac{du}{dr} - \frac{2}{r^2}u = 0 \tag{4}$$

The general solution of the above equation is

$$u = \bar{A}r + \frac{\bar{B}}{r^2} \tag{5}$$

Substituting Eq. (5) to Eq. (3) and using Eq. (2), we obtain the general elastic solution for stresses,

$$\sigma_r = A - \frac{2B}{r^3}, \qquad \sigma_\theta = A + \frac{B}{r^3} \tag{6}$$

where the constants A and B are related to the constants $\bar{A}$, $\bar{B}$ by

$$3K\bar{A} = A, \quad 2G\bar{B} = B$$

Using the boundary conditions

$$\sigma_r \,|_{r=a} = -p \tag{7}$$

$$\sigma_r \,|_{r=b} = 0 \tag{8}$$

we can determine the constants A and B as

$$A = \frac{a^3 p}{b^3 - a^3}, \qquad 2B = \frac{a^3 b^3 p}{b^3 - a^3}$$

and obtain the elastic solution for stresses and displacement as

$$\sigma_r = \frac{a^3 p}{b^3 - a^3}\left(1 - \frac{b^3}{r^3}\right) \le 0, \qquad \sigma_\phi = \sigma_\theta = \frac{a^3 p}{b^3 - a^3}\left(1 + \frac{b^3}{2r^3}\right) > 0 \tag{9}$$

$$u = \frac{a^3 p}{b^3 - a^3}\left[\frac{r}{3K} + \frac{1}{4G}\frac{b^3}{r^2}\right] \tag{10}$$

(a) The Tresca criterion has the form

$$\sigma_\theta - \sigma_r = \sigma_0 \tag{11}$$

where σ_0 is the tensile yield strength. Yielding begins at $r = a$, and the elastic limit pressure is

$$p_e = \frac{2\sigma_0}{3}\left(1 - \frac{a^3}{b^3}\right) \tag{12}$$

(b) For $p > p_e$, assuming the elastic-plastic boundary at $r = c$. To obtain the solution in the plastic region, we substitute the yield condition Eq. (11) to Eq. (1), and obtain,

$$\frac{d\sigma_r^{(1)}}{dr} + \frac{2\sigma_0}{r} = 0$$

where the sup-script (1) denotes the stress components in the plastic region. Solve the above equation, and obtain the general solution as

$$\sigma_r^{(1)} = 2\sigma_0 \, ln \, r + D$$

Use the boundary condition (7), and determine the integration constant D. The solution in the plastic region can therefore be expressed as

$$\sigma_r^{(1)} = -p + 2\sigma_0 \, ln \, \frac{r}{a}, \qquad \sigma_\theta^{(1)} = -p + \sigma_0 \, (1 + 2 \, ln \, \frac{r}{a}) \tag{13}$$

In the elastic region, the general elastic solution, Eq. (6), remains valid. Using the boundary condition, Eq. (8), we obtain

$$\sigma_r^{(2)} = 2B \, (\frac{1}{b^3} - \frac{1}{r^3}), \qquad \sigma_\theta^{(2)} = 2B \, (\frac{1}{b^3} + \frac{1}{2r^3})$$

where the sup-script (2) denotes the stress components in the elastic region.

To determine the constant B and derive the relationship between c and the internal pressure p, we use the continuity conditions at r = c,

$$\sigma_r^{(2)} \,|_{r=c} = \sigma_r^{(1)} \,|_{r=c}, \qquad (\sigma_\theta^{(2)} - \sigma_r^{(2)}) \,|_{r=c} = \sigma_0$$

and obtain

$$B = \frac{c^3}{3} \, \sigma_0 = \frac{b^3 c^3}{2(b^3 - c^3)} \, (p - 2\sigma_0 \, ln \, \frac{c}{a}) \tag{15}$$

$$p = 2\sigma_0 \, ln \, \frac{c}{a} + \frac{2}{3} \, \sigma_0 \, (1 - \frac{c^3}{b^3}) \tag{16}$$

While the stress components in the elastic region are

$$\sigma_r = \frac{2c^3}{3} \, \sigma_0 \, (\frac{1}{b^3} - \frac{1}{r^3}), \qquad \sigma_\theta = \frac{2c^3}{3} \, \sigma_0 \, (\frac{1}{b^3} + \frac{1}{2r^3})$$

and the plastic limit pressure can therefore be obtained by simply replacing c by b in Eq. (16),

$$p_c = 2 \, \sigma_0 \, ln \, \frac{b}{a} \tag{18}$$

Prob. 5.34 Same thick-walled sphere problem as in Prob. 5.33. Use the Rankine criterion.

Answer: $p_e = \frac{b^3 - a^3}{b^3 + a^3} \, \sigma_t, \quad p_c = (\frac{b^2}{a^2} - 1) \, \sigma_t.$

Prob. 5.35 Same thick-walled sphere problem as in Prob. 5.33. Use the Mohr-Coulomb criterion.

Answer:

$$p_e = \frac{b^3 - a^3}{m\,(a^3 + b^3) + (b^3 - a^3)}\,\sigma_c$$

$$p_c = \frac{\sigma_c}{m-1}\,[\,1 + (\frac{a}{b})^{-n}\,], \quad n = \frac{2\,(m-1)}{m}$$

5.8 Incremental Stress-Strain Relation Problems

Prob. 5.36 Derive the expression of the scalar factor $d\lambda$ for a general elastic-perfectly plastic material using the associated flow rule

$$d\varepsilon_{ij}^{p} = d\lambda\,\frac{\partial f}{\partial \sigma_{ij}}$$

where $f = f(\sigma_{ij})$ is the yield function. Assuming the elastic behavior is linear and isotropic, express the scalar factor $d\lambda$ in terms of the two elastic constants K and G.

Solution: Using the consistency condition

$$df = \frac{\partial f}{\partial \sigma_{ij}}\,d\sigma_{ij} = 0$$

and substituting

$$d\sigma_{ij} = C_{ijkl}\,(d\varepsilon_{kl} - d\varepsilon_{kl}^{p}) = C_{ijkl}\,(d\varepsilon_{kl} - d\lambda\,\frac{\partial f}{\partial \sigma_{kl}})$$

we obtain

$$\frac{\partial f}{\partial \sigma_{ij}}\,C_{ijkl}\,d\varepsilon_{kl} = d\lambda\,\frac{\partial f}{\partial \sigma_{ij}}\,C_{ijkl}\,\frac{\partial f}{\partial \sigma_{kl}}$$

or

$$d\lambda = \frac{C_{ijkl}\,\dfrac{\partial f}{\partial \sigma_{ij}}\,d\varepsilon_{kl}}{C_{lmrs}\,\dfrac{\partial f}{\partial \sigma_{lm}}\,\dfrac{\partial f}{\partial \sigma_{rs}}}$$

For a linear elastic isotropic material, we have

$$C_{ijkl} = \lambda\,\delta_{ij}\delta_{kl} + \mu\,(\delta_{ik}\delta_{jl} + \delta_{il}\delta_{jk})$$

$$= (K - \frac{2}{3}\,G)\,\delta_{ij}\delta_{kl} + G(\delta_{ik}\delta_{jl} + \delta_{il}\delta_{jk})$$

It follows that

$$C_{ijkl} \frac{\partial f}{\partial \sigma_{ij}} = (K - \frac{2}{3} G) (\frac{\partial f}{\partial \sigma_{ij}} \delta_{ij}) \delta_{kl} + G (\frac{\partial f}{\partial \sigma_{kl}} + \frac{\partial f}{\partial \sigma_{kl}})$$

$$= 2G [\frac{\partial f}{\partial \sigma_{kl}} + \frac{3K - 2G}{6G} (\frac{\partial f}{\partial \sigma_{ij}} \delta_{ij}) \delta_{kl}]$$

$$C_{ijkl} \frac{\partial f}{\partial \sigma_{ij}} \frac{\partial f}{\partial \sigma_{kl}} = 2G [\frac{\partial f}{\partial \sigma_{kl}} \frac{\partial f}{\partial \sigma_{kl}} + \frac{3K-2G}{6G} (\frac{\partial f}{\partial \sigma_{ij}} \delta_{ij}) \frac{\partial f}{\partial \sigma_{kl}} \delta_{kl}]$$

$$= 2G [\frac{\partial f}{\partial \sigma_{ij}} \frac{\partial f}{\partial \sigma_{ij}} + \frac{3K-2G}{6G} (\frac{\partial f}{\partial \sigma_{ij}} \delta_{ij})^2]$$

$$C_{ijkl} \frac{\partial f}{\partial \sigma_{ij}} d\varepsilon_{kl} = 2G [\frac{\partial f}{\partial \sigma_{kl}} d\varepsilon_{kl} + \frac{3K-2G}{6G} (\frac{\partial f}{\partial \sigma_{ij}} \delta_{ij}) d\varepsilon_{kk}]$$

$$= 2G [\frac{\partial f}{\partial \sigma_{ij}} d\varepsilon_{ij} + \frac{3K-2G}{6G} (\frac{\partial f}{\partial \sigma_{ij}} \delta_{ij}) d\varepsilon_{kk}]$$

Therefore, we obtain

$$d\lambda = \frac{\frac{\partial f}{\partial \sigma_{ij}} d\varepsilon_{ij} + \frac{3K-2G}{6G} (\frac{\partial f}{\partial \sigma_{ij}} \delta_{ij}) d\varepsilon_{kk}}{\frac{\partial f}{\partial \sigma_{lm}} \frac{\partial f}{\partial \sigma_{lm}} + \frac{3K-2G}{6G} (\frac{\partial f}{\partial \sigma_{rs}} \delta_{rs})^2}$$

Prob. 5.37 Show that for a perfectly-plastic isotropic material with a non-associated flow rule, the incremental stress-strain relation has the form

$$d\sigma_{ij} = \left[C_{ijkl} - \frac{C_{ijmn} \frac{\partial f}{\partial \sigma_{mn}} \frac{\partial g}{\partial \sigma_{pq}} C_{pqkl}}{\frac{\partial f}{\partial \sigma_{rs}} C_{rstu} \frac{\partial g}{\partial \sigma_{tu}}} \right] d\varepsilon_{kl}$$

where f is the yield function, and g the plastic potential function.

Prob. 5.38 Derive the incremental stress-strain relation for an isotropic perfectly-plastic material with a non-associated flow rule. Use the Drucker-Prager criterion as the yield condition, and the von Mises function as the plastic potential.

Answer:

$$C_{ijkl}^{ep} = C_{ijkl} - \frac{1}{\sqrt{J_2}} (3K\alpha\delta_{kl} + \frac{G}{\sqrt{J_2}} s_{kl})\, s_{ij}$$

Prob. 5.39 Derive the incremental stress-strain relation for an isotropic perfectly-plastic material with a non-associated flow rule. Use the yield function

$$f(\sigma_{ij}) = \alpha I_1 + J_2^{1/2} - k = 0$$

and the plastic potential function

$$g(\sigma_{ij}) = \bar{\alpha} I_1 + J_2^{1/2}$$

Answer:

$$C_{ijkl}^{ep} = C_{ijkl} - \frac{1}{G + 9K\alpha\bar{\alpha}} [3K\alpha\delta_{kl} + \frac{G}{\sqrt{J_2}} s_{kl}]\, [3K\bar{\alpha}\delta_{ij} + \frac{G}{\sqrt{J_2}} s_{ij}]$$

5.9 Source Code Listing of PLASTIC_ZONE

The program PLASTIC_ZONE is written in C language. It consists of only one file, *plastic_zone.c*. The following is the source code listing of PLASTIC1 in its entirety.

```
/*
 * PLASTIC_ZONE
 *
 * Purpose: Compute the shape of the near tip plastic zone for a
 *          general mixed mode crack
 *
 * Input  : 1) type of criterion
 *            "Mises"          for von Mises criterion
 *            "Tresca"         for Tresca criterion
 *            "Rankine"        for Rankine criterion
 *            "Mohr"           for Mohr-Coulomb criterion
 *            "Drucker"        for Drucker-Prager criterion
 *          2) type of deformation
 *            "stress"         for plane stress
 *            "strain"         for plane strain
 *          3) base name of the output file
 *          4) (k1)min, (k1)max, and (k1)inc
 *            Note: 0 < k1 < 1, and the increment of k1 must larger
 *                  than 0
 *                k2 = sqrt(1 - k1*k1)
 *          5) yielding parameter
 *            for Mohr-Coulomb and Drucker-Prager Criteria only
 *                for Mohr-Coulomb criterion        : m
 *                for the Drucker-Prager criterion : alpha
 *
 * Output : for each k1, the x, y coord. of the plastic zone shape
 *          are computed and stored in a file with a name of
 *          "basenameX", where the "basename" is inputed as the
 *          third data, and X is a number starting from 0.
 *
 * Compile: cc plastic_zone.c -lm
 *
 */
#include <stdio.h>
#include <math.h>
#include <string.h>

#define MAX(a, b)     ((a) > (b) ? (a) : (b))
#define MIN(a, b)     ((a) < (b) ? (a) : (b))
#define ABS(a)        ((a) < 0  ? -(a) : (a))
#define SIGN(a)       ((a) >= 0  ?  1  : -1 )

#ifndef FALSE
#define TRUE          1
#define FALSE         0
#endif
```

```
#define FXX(t)  (cos(t/2.0) * (1.0 - sin(t/2.0) * sin(t*3.0/2.0)))
#define FYY(t)  (cos(t/2.0) * (1.0 + sin(t/2.0) * sin(t*3.0/2.0)))
#define FXY(t)  (cos(t/2.0) * sin(t/2.0) * cos(t*3.0/2.0))

#define GXX(t)  (- sin(t/2.0) * (2.0 + cos(t/2.0) * cos(t*3.0/2.0)))
#define GYY(t)  (sin(t/2.0) * cos(t/2.0) * cos(t*3.0/2.0))
#define GXY(t)  (cos(t/2.0) * (1.0 - sin(t/2.0) * sin(t*3.0/2.0)))

#define PI        3.1415926
#define NU        0.3
#define MAXPOINTS 121
#define NUF1      (1.0 - NU + NU*NU)
#define NUF2      (1.0 + 2.0*NU - 2.0*NU*NU)

static int model_type = -1, plane_stress = FALSE;
double Mises(), Tresca(), Rankine(), Mohr(), Drucker();
static double (*sbCriteria[])() = { Mises,
                                    Tresca,
                                    Rankine,
                                    Mohr,
                                    Drucker  };
static char *sbMnames[] = { "Mises",
                            "Tresca",
                            "Rankine",
                            "Mohr",
                            "Drucker"  };

/*
 * find the stress invariants
 */
static void
Invariants(pI1, pJ2, Sxx, Syy, Sxy)
double *pI1, *pJ2;
double Sxx, Syy, Sxy;
{
    *pI1 = (plane_stress) ? (Sxx + Syy) : (1.0 + NU)*(Sxx + Syy);
    *pJ2 = (plane_stress) ?
        ((Sxx*Sxx + Syy*Syy - Sxx*Syy)/3.0 + Sxy*Sxy) :
        ((NUF1*(Sxx*Sxx + Syy*Syy) - NUF2*Sxx*Syy)/3.0 + Sxy*Sxy);
    return;
}

/*
 * find the principal stresses
 */
static void
Principals(pS1, pS2, pS3, Sxx, Syy, Sxy)
double *pS1, *pS2, *pS3;
double Sxx, Syy, Sxy;
{
    auto double Szz, p, q, c, S11, S22;
```

```
    Szz = (plane_stress) ? 0.0 : NU*(Sxx + Syy);
    p = (Sxx + Syy)/2.0;
    c = Sxx - Syy;
    q = sqrt(c*c/4.0 + Sxy*Sxy);
    S11 = p + q;
    S22 = p - q;
    *pS1 = MAX(Szz, MAX(S11, S22));
    *pS3 = MIN(Szz, MIN(S11, S22));
    *pS2 = Sxx + Syy + Szz - *pS1 - *pS3;
    return;
}

/*
 * for the Mises criterion
 */
static double
Mises(no_use, Sxx, Syy, Sxy)
double no_use, Sxx, Syy, Sxy;
{
    auto double I1, J2;

    Invariants(&I1, &J2, Sxx, Syy, Sxy);
    return (3.0*J2);
}

/*
 * for the Tresca criterion
 */
static double
Tresca(no_use, Sxx, Syy, Sxy)
double no_use, Sxx, Syy, Sxy;
{
    auto double tau, S1, S2, S3;

    Principals(&S1, &S2, &S3, Sxx, Syy, Sxy);
    tau = S1 - S3;
    return (tau*tau);
}

/*
 * for the Rankine criterion
 */
static double
Rankine(no_use, Sxx, Syy, Sxy)
double no_use, Sxx, Syy, Sxy;
{
    auto double S1, S2, S3;

    Principals(&S1, &S2, &S3, Sxx, Syy, Sxy);
    return (S1 > 0.0) ? (S1*S1) : 0.0;
}
```

```
/*
 * for the Mohr-Coulomb criterion
 */
static double
Mohr(m, Sxx, Syy, Sxy)
double m, Sxx, Syy, Sxy;
{
    auto double tau, S1, S2, S3;

    Principals(&S1, &S2, &S3, Sxx, Syy, Sxy);
    tau = (m*S1 - S3) / m;
    return (tau*tau);
}

/*
 * for the Drucker-Prager criterion
 */
#define SQRT3I        0.57735027
static double
Drucker(alpha, Sxx, Syy, Sxy)
double alpha, Sxx, Syy, Sxy;
{
    auto double I1, J2, c;

    Invariants(&I1, &J2, Sxx, Syy, Sxy);
    c = (alpha * I1 + sqrt(J2)) / (alpha + SQRT3I);
    return (c*c);
}

int
main()
{
    auto double theta, dtheta;
    auto double s1, s2, s10, s11, ds, para;
    auto double Sxx, Syy, Sxy, S11, S22, S33, I1, J2;
    auto double rp, sbX[MAXPOINTS], sbY[MAXPOINTS];
    auto char sbFname[20], sbBaseName[15], sbModelName[10];
    auto char sbDeform[10];
    auto char sbNumber[3];
    auto FILE *pF;
    auto int i, case_count, n;

    dtheta = 2.0*PI/((double) MAXPOINTS - 1.0);
    /* get input    */
    scanf("%s", sbModelName);
    for (i = 0; i < 5; ++i)
        if (0 == strcmp(sbModelName, sbMnames[i]))
            model_type = i;
    if (-1 == model_type) {
        printf("Unknown model name : \"%s\". Stop.\n", sbModelName);
        exit(2);
    } else {
```

```
        printf("Use model \"%s\"\n", sbMnames[model_type]);
    }
    scanf("%s", sbDeform);
    if (0 == strcmp(sbDeform, "stress"))
        plane_stress = TRUE;
    printf("Plane %s condition.\n",
          (plane_stress) ? "stress" : "strain");
    scanf("%s", sbBaseName);
    printf("Base name of the output files : \"%s\"\n", sbBaseName);
    scanf("%lf %lf %lf", &s10, &s11, &ds);
    case_count = (int) ((s11 - s10) / ds) + 1;
    printf("Total %d cases.\n", case_count);
    printf("s1 initial\t: %f\ns1 final\t: %f\ns1 increment\t: %f\n",
        s10, s11, ds);
    if (model_type > 2) {
        scanf("%lf", &para);
        printf("Parameter in calculation\t: %f\n", para);
    }
    for (n = 0, s1 = s10; n < case_count; ++n, s1+=ds) {
        s2 = sqrt(1.0 - s1*s1);
        for (i = 0, theta= -PI; i < MAXPOINTS; ++i, theta+=dtheta) {
            Sxx = s1 * FXX(theta) + s2 * GXX(theta);
            Syy = s1 * FYY(theta) + s2 * GYY(theta);
            Sxy = s1 * FXY(theta) + s2 * GXY(theta);
            rp = sbCriteria[model_type](para, Sxx, Syy, Sxy);
            sbX[i] = rp * cos(theta);
            sbY[i] = rp * sin(theta);
        }
        strcpy(sbFname, sbBaseName);
        sprintf(sbNumber, "%d", n);
        strcat(sbFname, sbNumber);
        pF = fopen(sbFname, "w");
        for (i = 0; i < MAXPOINTS; ++i)
            fprintf(pF, "%f\t%f\n", sbX[i], sbY[i]);
        fclose(pF);
    }
    exit(0);
}
```

Chapter 6

Hardening Plastic Stress Analysis

6.1 Introduction

Most engineering materials exhibit work-hardening behavior. Increasing a stress state beyond the initial yield surface and into the work-hardening range results in both elastic and plastic deformations. With the development of plastic deformation, a new yield surface, also called *subsequent yield or loading surface*, evolves. If the stress point moves inside the subsequent loading surface, the material behaves elastically. and no plastic deformation takes place. Since loading surfaces change according to the way the plastic deformation was produced, the stress-strain behavior of a work-hardening material is *loading path dependent* or *loading history dependent.*

There are two types of theory dealing with work-hardening materials: the *deformational theory* and the *incremental theory.* The deformational theory neglects the loading history dependency in the development of stress-strain relationships by assuming that the stress state, σ_{ij}, determines uniquely the total strain and plastic strain state, ε_{ij} and ε_{ij}^{p}, as long as plastic deformation continues. Because of its relatively simplicity, the deformational theory has been used extensively in engineering practice for solving elastic-plastic problems. The general validity of the theory is limited to the monotonically increasing loading in which: (1) the stress components are increased nearly proportionally in a loading process, known as *proportional loading*; and (2) no unloading occurs.

In contrast to the deformational theory, loading path dependency is assured in the incremental theory. In this theory, the plastic strain increment, $d\varepsilon_{ij}^{p}$, is related to the stress increment, $d\sigma_{ij}$, through the plastic deformation history, κ, and the current stress state, σ_{ij}. The theory of perfect plasticity described in the preceding chapter is the simplest type of incremental theory. Techniques such as flow rules used in the perfectly-plastic analysis can be extended to work-hardening analysis with little changes. The fundamental difference between perfect and hardening plasticity is that the yield surface is no longer fixed in the stress space for a hardening material. The rule governing the changes of subsequent yield surfaces is called the *hardening rule*. There are three basic elements in the development of work-hardening stress-strain relationships: (1) loading criteria; (2) flow rules; and (3) hardening rules.

In this chapter, the highlights of the deformational theory are described first in Section 6.2, followed by a brief summary of the incremental theory in Sections 6.3 to 6.7. Problems associated with the incremental theory and the deformational theory are then presented in Sections 6.8 and 6.9 respectively.

6.2 Deformational Theory

The simplest deformational theory is the J_2-*deformational theory* based on the von Mises yield criterion. The total strain ε_{ij} is decomposed into the elastic and plastic components as

$$\varepsilon_{ij} = \varepsilon^e_{ij} + \varepsilon^p_{ij} \tag{6.1}$$

The elastic strain component ε^e_{ij} is related to the stress σ_{ij} by the Hooke's law

$$\varepsilon^e_{ij} = \frac{s_{ij}}{2G} + \frac{\sigma_{kk}}{9K}\,\delta_{ij} \tag{6.2}$$

and the plastic strain ε^p_{ij} component is related to the deviatoric stress s_{ij} as

$$\varepsilon^p_{ij} = \phi\, s_{ij}, \quad \phi = \phi(J_2) \tag{6.3}$$

where ϕ is a scalar function of the invariant J_2. To calibrate the function $\phi(J_2)$ with experimental stress-strain curves, we introduce the stress variable, called *effective stress* σ_e, defined as

$$\sigma_e = \sqrt{3\,J_2} \tag{6.4}$$

and the plastic strain variable, called *effective plastic strain* ε_p, defined as

$$\varepsilon_p = \sqrt{\frac{2}{3}\,\varepsilon^p_{ij}\varepsilon^p_{ij}} \tag{6.5}$$

These two variables can be related to each other by the stress-plastic strain relationship of material under uniaxial tension test

$$\sigma_e = \sigma_e(\varepsilon_p) \tag{6.6}$$

From which the function ϕ is determined as

$$\phi = \frac{3}{2}\frac{\varepsilon_p}{\sigma_e} = \frac{\sqrt{3}}{2}\frac{\varepsilon_p}{\sqrt{J_2}} \tag{6.7}$$

Using the function ϕ, plastic strain increments can now be expressed explicitly in component form as

$$\varepsilon_x^p = \frac{\varepsilon_p}{\sigma_e}\left[\sigma_x - \frac{1}{2}(\sigma_y + \sigma_z)\right], \quad \gamma_{xz} = \frac{3\,\varepsilon_p}{\sigma_e}\tau_{xz} \tag{6.8a}$$

$$\varepsilon_y^p = \frac{\varepsilon_p}{\sigma_e}\left[\sigma_y - \frac{1}{2}(\sigma_z + \sigma_x)\right], \quad \gamma_{yz} = \frac{3\,\varepsilon_p}{\sigma_e}\tau_{yz} \tag{6.8b}$$

$$\varepsilon_z^p = \frac{\varepsilon_p}{\sigma_e}\left[\sigma_z - \frac{1}{2}(\sigma_x + \sigma_y)\right], \quad \gamma_{xy} = \frac{3\,\varepsilon_p}{\sigma_e}\tau_{xy} \tag{6.8c}$$

For a given stress state in a plastic loading process, the effective stress σ_e can be determined from Eq. (6.4). The effective plastic strain ε_p can then be obtained from Eq. (6.6). Thus, the complete plastic strain components are obtained from Eq. (6.8)

6.3 Loading Surfaces and Loading Criteria

A *loading surface* is a subsequent yield surface that defines the new boundary of the current elastic region. A loading surface may be generally expressed as a function of the current stress state, σ_{ij}, and the accumulated plastic deformation represented by ε_{ij}^p, and the hardening parameter κ as

$$f(\sigma_{ij}, \varepsilon_{ij}^p, \kappa) = 0 \tag{6.9}$$

The loading surface is related to the initial yield surface, f_0, as

$$f_0(\sigma_{ij}) = f(\sigma_{ij}, 0, 0)$$

For stress points moving on the loading surface or inward, no incremental plastic deformation occurs (i.e., $d\varepsilon_{ij}^p = 0$). For stress points moving outward of the surface, incremental plastic deformation occurs and a new yield surface evolves. For a stress point on the current loading surface, i.e., $f(\sigma_{ij}, \varepsilon_{ij}^p, \kappa) = 0$, the condition determining whether a further stress increment or strain increment will cause a plastic loading or not is called *loading criterion*. In terms of stress increment, the loading criterion can be expressed in the form as

$$\textit{loading:} \qquad \frac{\partial f}{\partial \sigma_{ij}}\, d\sigma_{ij} > 0 \tag{6.10a}$$

$$\textit{neutral loading:} \qquad \frac{\partial f}{\partial \sigma_{ij}}\, d\sigma_{ij} = 0 \tag{6.10b}$$

unloading: $$\frac{\partial f}{\partial \sigma_{ij}} d\sigma_{ij} < 0 \tag{6.10c}$$

In the case of loading, the stress point moves outward of the loading surface and further plastic deformation occurs, $d\varepsilon_{ij}^p \neq 0$. In the case of unloading, the stress point moves inward and $d\varepsilon_{ij}^p = 0$. In the case of neutral loading, the stress point moves on the loading surface, and no further plastic deformation occurs, $d\varepsilon_{ij}^p = 0$.

In terms of strain increment, the loading criterion can be written as

loading: $$\frac{\partial f}{\partial \sigma_{ij}} C_{ijkl}\, d\varepsilon_{kl} > 0 \tag{6.11a}$$

neutral loading: $$\frac{\partial f}{\partial \sigma_{ij}} C_{ijkl}\, d\varepsilon_{kl} = 0 \tag{6.11b}$$

unloading: $$\frac{\partial f}{\partial \sigma_{ij}} C_{ijkl}\, d\varepsilon_{kl} < 0 \tag{6.11c}$$

where C_{ijkl} is the elastic stiffness tensor.

6.4 Flow Rules

The plastic strain increment vector in strain space is in the direction of the gradient of a *plastic potential function* g, similar to that of the perfectly-plastic materials discussed in the preceding chapter as

$$d\varepsilon_{ij}^p = d\lambda \frac{\partial g}{\partial \sigma_{ij}} \tag{6.12}$$

Here, as in the subsequent yield surface $f(\sigma_{ij}, \varepsilon_{ij}^p, \kappa)$, the potential function g depends not only on stress state, but also on accumulated plastic deformations as

$$g = g(\sigma_{ij}, \varepsilon_{ij}^p, \kappa) \tag{6.13}$$

Equation (6.12) is called *non-associated flow rule* for work-hardening materials. It reduces to the *associated flow rule* case when the subsequent yield function f is selected as the potential function g

$$d\varepsilon_{ij}^p = d\lambda \frac{\partial f}{\partial \sigma_{ij}} \tag{6.14}$$

6.5 Hardening Rules

A work-hardening rule defines the way a new loading surface evolves according to the development of plastic deformations. There are three widely used hardening rules, among others: *isotropic hardening*, *kinematic hardening*, and *mixed hardening*.

The general form of the mixed hardening can be written simply as

$$f(\sigma_{ij}, \varepsilon_{ij}^{p}, \kappa) = F(\sigma_{ij}, \varepsilon_{ij}) - k^2(\kappa) = F(\sigma_{ij} - \alpha_{ij}) - k^2(\kappa) = 0 \qquad (6.15)$$

where function $k^2(\kappa)$ represents the size of the loading surface, and k^2 is used here to emphasize that it is always positive; the function F defines the shape of the surface, and α_{ij} denotes the coordinate of the center of the loading surface in stress space.

6.5.1 Isotropic Hardening

The isotropic hardening rule assumes that the subsequent loading surfaces expand uniformly from its initial yield surface without distortion and translation as plastic deformation develops. The size of the subsequent loading surfaces is controlled by the hardening parameter κ through the function $k^2(\kappa)$

$$f(\sigma_{ij}, \varepsilon_{ij}, \kappa) = F(\sigma_{ij}) - k^2(\kappa) = 0 \qquad (6.16)$$

For example, the isotropic hardening von Mises loading surface can be written in the simple form

$$f(\sigma_{ij}, \varepsilon_{ij}, \kappa) = J_2 - \frac{1}{3}\sigma_e^2(\kappa) \qquad (6.17)$$

where

$$F(\sigma_{ij}) = J_2, \qquad k^2 = \frac{1}{3}\sigma_e^2$$

σ_e is the effective stress. This will be discussed further in the next section.

6.5.2 Kinematic Hardening

The kinematic hardening rule assumes that the subsequent loading surface translates as a rigid body in stress space as plastic deformation develops without changing the size, shape, and orientation of the initial yield surface. This hardening rule provides a simple way to account for the Bauschinger effect. The translation of the loading surface is governed by the plastic strain ε_{ij}^{p} through the center coordinate α_{ij} of the loading surface

$$f(\sigma_{ij}, \varepsilon_{ij}, \kappa) = F(\sigma_{ij} - \alpha_{ij}) - k^2 = 0 \tag{6.18}$$

For example, the kinematic hardening von Mises loading surface can be written in the simple form

$$f(\sigma_{ij}, \varepsilon_{ij}, \kappa) = \frac{1}{2}(s_{ij} - \alpha_{ij})(s_{ij} - \alpha_{ij}) - \frac{1}{3}\sigma_0^2 \tag{6.19}$$

in which we have used

$$F(\sigma_{ij} - \alpha_{ij}) = \frac{1}{2}(s_{ij} - \alpha_{ij})(s_{ij} - \alpha_{ij})$$

$$k^2 = \frac{1}{3}\sigma_0^2 = \text{constant}$$

in Eq. (6.18), and σ_0 is the uniaxial tensile strength.

The center coordinate of the loading surface, α_{ij}, can be related to the plastic strain increment by either the Prager's rule

$$d\alpha_{ij} = c\, d\varepsilon_{ij}^p, \quad \text{or} \quad \alpha_{ij} = c\, \varepsilon_{ij}^p \tag{6.20}$$

where $c > 0$ is a material constant; or the Ziegler's rule

$$d\alpha_{ij} = d\mu\,(\sigma_{ij} - \alpha_{ij}), \qquad d\mu = a\, d\varepsilon_p \tag{6.21}$$

where $a > 0$ is a material constant, and ε_p is the effective plastic strain. This will be discussed further in the next section.

6.5.3 Mixed Hardening

The mixed hardening rule, Eq. (6.15), combines the behaviors of both isotropic and kinematic hardenings. The subsequent loading surface, as plastic deformation develops, experiences a translation defined by α_{ij} and a uniform expansion measured by $k^2(\kappa)$. However, the original shape and orientation of the initial yield surface remains the same. In this rule, different degrees of Bauschinger effect can be simulated by simply adjusting α_{ij} and k^2.

For example, the mixed hardening von Mises surface can be expressed as

$$f(\sigma_{ij}, \varepsilon_{ij}, \kappa) = \frac{1}{2}(s_{ij} - \alpha_{ij})(s_{ij} - \alpha_{ij}) - \frac{1}{3}\sigma_e^2(\kappa) \tag{6.22}$$

6.6 Effective Stress and Effective Plastic Strain

To apply the work-hardening theory to practical engineering analysis, we must relate the function $k^2(\kappa)$ and the hardening parameter κ to stress and plastic strain variables such that they can be determined from experimental simple uniaxial stress-strain curve. To this end, a stress variable σ_e, called *effective stress*, and a plastic strain variable ε_p, called *effective plastic strain*, are introduced. The *effective stress-effective plastic strain curve* can then be used to calibrate the k^2 and κ with the uniaxial experimental stress-strain curve. The effective stress-effective plastic strain curve reduces to a uniaxial stress-strain curve under a uniaxial stress condition.

6.6.1 Effective Stress

For an isotropic hardening material, the function $F(\sigma_{ij})$ defined in Eq. (6.16) can be used to define the effective stress. Since $F(\sigma_{ij})$ is a homogeneous function of stress components of degree n, and we need to reduce the effective stress σ_e to the uniaxial stress σ_1 under uniaxial stress condition, it follows that the function F and the effective stress σ_e are related by

$$F(\sigma_{ij}) = C\,\sigma_e^n \tag{6.23}$$

where C and n are constants. They can be uniquely determined by the selection of a loading surface. For example, for the von Mises loading surface, $F(\sigma_{ij}) = J_2$ with the uniaxial stress condition, $J_2 = \frac{1}{3}\sigma_1^2$, we determine

$$n = 2, \quad C = \frac{1}{3}, \quad \sigma_e = \sqrt{3\,J_2}$$

For a kinematic hardening material, we denote

$$\bar{\sigma}_{ij} = \sigma_{ij} - \alpha_{ij} \tag{6.24}$$

as the *reduced stresses* which are associated only with the expansion of loading surface, the effective stress, called *reduced effective stress*, is defined as, similar to Eq. (6.23) as

$$F(\sigma_{ij} - \alpha_{ij}) = F(\bar{\sigma}_{ij}) = C\,\bar{\sigma}_e^n \tag{6.25}$$

6.6.2 Effective Plastic Strain

We first define the effective plastic strain ε_p and then use it later as the hardening parameter κ. The effective plastic strain can be defined in terms of plastic work per unit volume, dW_p, through the effective stress σ_e as

$$dW_p = \sigma_e \, d\varepsilon_p \tag{6.26}$$

By definition, the plastic work increment can be written as

$$dW_p = \sigma_{ij} \, d\varepsilon_{ij}^p = d\lambda \, \sigma_{ij} \frac{\partial f}{\partial \sigma_{ij}} = d\lambda \, \sigma_{ij} \frac{\partial F}{\partial \sigma_{ij}} \tag{6.27}$$

Combining Equations (6.26) and (6.27) and noting the fact that F is a homogeneous function of degree n in stresses

$$\sigma_{ij} \frac{\partial F}{\partial \sigma_{ij}} = n F$$

and

$$d\lambda = \frac{\sqrt{d\varepsilon_{ij}^p \, d\varepsilon_{ij}^p}}{\sqrt{\dfrac{\partial F}{\partial \sigma_{kl}} \dfrac{\partial F}{\partial \sigma_{kl}}}}$$

we obtain

$$dW_p = d\lambda \, n F = \frac{\sqrt{d\varepsilon_{ij}^p \, d\varepsilon_{ij}^p} \, n F}{\sqrt{\dfrac{\partial F}{\partial \sigma_{kl}} \dfrac{\partial F}{\partial \sigma_{kl}}}} = \sigma_e \, d\varepsilon_p \tag{6.28}$$

Equation (6.28) can be used to define the increment of effective plastic strain $d\varepsilon_p$. For example, for the von Mises surface, $n = 2$, $F = J_2$, and $\partial F / \partial \sigma_{ij} = s_{ij}$, we find

$$d\varepsilon_p = \sqrt{\frac{2}{3} \, d\varepsilon_{ij}^p \, d\varepsilon_{ij}^p}$$

6.6.3 Effective Stress-Effective Plastic Strain Relation

Since the effective stress and effective plastic strain can both be reduced to the uniaxial stress σ_1 and uniaxial plastic strain ε_1^p respectively under uniaxial stress condition, They can therefore be calibrated against the uniaxial stress-plastic strain test curve

$$\sigma_e = \sigma_e(\varepsilon_p) \tag{6.29}$$

Differentiation of Equation (6.29) leads to

$$d\sigma_e = H_p(\sigma_e)\, d\varepsilon_p, \quad \text{or} \quad H_p = \frac{d\sigma_e}{d\varepsilon_p} \tag{6.30}$$

where H_p is the *plastic modulus* associated with the rate of expansion of the loading surface which can be considered here as the slope of the uniaxial stress-plastic strain curve at the current value σ_e.

6.7 Incremental Stress-Strain Relationships

The incremental stress-strain relation for an elastic-work-hardening plastic material is derived here. The yield surface or loading surface of a work-hardening material has the general form

$$f(\sigma_{ij}, \varepsilon_{ij}^p, \kappa) = 0 \tag{6.31}$$

The total strain increment is decomposed into two parts,

$$d\varepsilon_{ij} = d\varepsilon_{ij}^e + d\varepsilon_{ij}^p \tag{6.32}$$

The elastic strain increment $d\varepsilon_{ij}^e$ is related to the stress increment by the Hooke's law as

$$d\sigma_{ij} = C_{ijkl}\, d\varepsilon_{kl}^e = C_{ijkl}\, (d\varepsilon_{kl} - d\varepsilon_{kl}^p) \tag{6.33}$$

while the plastic strain increment $d\varepsilon_{ij}^p$ can be expressed by a non-associated flow rule in the general form

$$d\varepsilon_{ij} = d\lambda \frac{\partial g}{\partial \sigma_{ij}} \tag{6.34}$$

where $g = g(\sigma_{ij}, \varepsilon_{ij}^p, \kappa)$ is the plastic potential function, and $d\lambda$ is a positive scalar to be determined. Since the current stress point must remain on the current loading surface during plastic loading, it must satisfy the consistency condition

$$df = \frac{\partial f}{\partial \sigma_{ij}} d\sigma_{ij} + \frac{\partial f}{\partial \varepsilon_{ij}^p} d\varepsilon_{ij}^p + \frac{\partial f}{\partial \kappa} d\kappa = 0 \tag{6.35}$$

The consistency condition imposes a restriction on the relationship among the three increments $d\sigma_{ij}$, $d\varepsilon_{ij}^p$ and $d\kappa$. Utilizing this condition, we can express the positive scalar $d\lambda$ and thus the plastic strain increment $d\varepsilon_{ij}^p$ in terms of either the stress increment $d\sigma_{ij}$ or the strain increment $d\varepsilon_{ij}$.

To this end, we shall first consider the last term in Eq. (6.35). The hardening parameter κ is here taken as the effective plastic strain ε_p as defined in the preceding section. The effective plastic strain increment can generally be expressed as

$$d\kappa = d\varepsilon_p = C\sqrt{d\varepsilon_{ij}^p d\varepsilon_{ij}^p} \tag{6.36}$$

where C is a loading-surface-related constant. Substituting the flow rule (6.34) into the above equation, we obtain

$$d\varepsilon_p = C\sqrt{\frac{\partial g}{\partial \sigma_{ij}}\frac{\partial g}{\partial \sigma_{ij}}}\, d\lambda \tag{6.37}$$

Substituting the flow rule (6.34) and Eq. (6.37) for $d\varepsilon_p$ into the consistency condition (6.35), we solve $d\lambda$ in terms of $d\sigma_{ij}$ as

$$d\lambda = \frac{1}{h}\frac{\partial f}{\partial \sigma_{ij}} d\sigma_{ij} \tag{6.38a}$$

where

$$h = -\frac{\partial f}{\partial \varepsilon_{ij}^p}\frac{\partial g}{\partial \sigma_{ij}} - \frac{\partial f}{\partial \varepsilon_p} C\sqrt{\frac{\partial g}{\partial \sigma_{ij}}\frac{\partial g}{\partial \sigma_{ij}}} \tag{6.38b}$$

From which we obtain the plastic strain increment in term of stress increment as

$$d\varepsilon_{ij}^p = \frac{1}{h}\frac{\partial g}{\partial \sigma_{ij}}\frac{\partial f}{\partial \sigma_{kl}} d\sigma_{kl} \tag{6.39}$$

Alternatively, substituting the flow rule (6.34), Eq. (6.33) for stress increment $d\sigma_{ij}$, and Eq. (6.37) for $d\varepsilon_p$ into the consistency condition (6.35), we can solve $d\lambda$ in terms of $d\varepsilon_{ij}$ as

$$d\lambda = \frac{1}{h}\frac{\partial f}{\partial \sigma_{ij}} C_{ijkl} d\varepsilon_{kl} \tag{6.40a}$$

where

$$h = \frac{\partial f}{\partial \sigma_{ij}} C_{ijkl}\frac{\partial g}{\partial \sigma_{kl}} - \frac{\partial f}{\partial \varepsilon_{ij}^p}\frac{\partial g}{\partial \sigma_{ij}} - \frac{\partial f}{\partial \varepsilon_p} C\sqrt{\frac{\partial g}{\partial \sigma_{ij}}\frac{\partial g}{\partial \sigma_{ij}}} \tag{6.40b}$$

Defining the following two second-order tensors

$$H_{kl} = \frac{\partial f}{\partial \sigma_{ij}} C_{ijkl}, \qquad H_{kl}^* = \frac{\partial g}{\partial \sigma_{ij}} C_{ijkl} \tag{6.41}$$

we can express the plastic strain increment $d\varepsilon_{ij}^p$ in term of the total strain increment $d\varepsilon_{ij}$ in the simple form

$$d\varepsilon_{ij}^p = \frac{1}{h}\frac{\partial g}{\partial \sigma_{ij}} H_{kl}\, d\varepsilon_{kl} \tag{6.42}$$

Using Eq. (6.42), Eq. (6.33) becomes

$$d\sigma_{ij} = [\, C_{ijkl} - \frac{1}{h} H^*_{ij} H_{kl} \,]\, d\varepsilon_{kl} = C^{ep}_{ijkl}\, d\varepsilon_{kl} \tag{6.43}$$

where C^{ep}_{ijkl} is the elastic-plastic tangential modulus.

The incremental stress-strain relation (6.43) is valid of course for plastic loading only. A complete incremental stress-strain relation is now summarized below.

Elastic loading: $f(\sigma_{ij}, \varepsilon^p_{ij}, \kappa) < 0$ (6.44a)

$$d\varepsilon^p_{ij} = 0, \qquad d\sigma_{ij} = C_{ijkl}\, d\varepsilon_{kl} \tag{6.44b}$$

Plastic loading: $f(\sigma_{ij}, \varepsilon^p_{ij}, \kappa) = 0, \quad \frac{\partial f}{\partial \sigma_{ij}} C_{ijkl}\, d\varepsilon_{kl} > 0$ (6.45a)

$$d\varepsilon^p_{ij} = \frac{1}{h} \frac{\partial g}{\partial \sigma_{ij}} H_{kl}\, d\varepsilon_{kl}, \qquad d\sigma_{ij} = C^{ep}_{ijkl}\, d\varepsilon_{kl} \tag{6.45b}$$

Neutral loading: $f(\sigma_{ij}, \varepsilon^p_{ij}, \kappa) = 0, \quad \frac{\partial f}{\partial \sigma_{ij}} C_{ijkl}\, d\varepsilon_{kl} = 0$ (6.46a)

$$d\varepsilon^p_{ij} = 0, \qquad d\sigma_{ij} = C_{ijkl}\, d\varepsilon_{kl} \tag{6.46b}$$

Unloading: $f(\sigma_{ij}, \varepsilon^p_{ij}, \kappa) = 0, \quad \frac{\partial f}{\partial \sigma_{ij}} C_{ijkl}\, d\varepsilon_{kl} < 0$ (6.47a)

$$d\varepsilon^p_{ij} = 0, \qquad d\sigma_{ij} = C_{ijkl}\, d\varepsilon_{kl} \tag{6.47b}$$

6.8 Problems Using Incremental Theory

Prob. 6.1 Show that for the von Mises loading surface with the isotropic hardening rule

$$f(\sigma_{ij}, \varepsilon^p_{ij}, \varepsilon_p) = J_2 - \frac{1}{3} \sigma_e^2(\kappa) = 0$$

the plastic strain increment can be expressed in term of strain increment as

$$d\varepsilon^p_{ij} = \frac{2G}{h} s_{ij} s_{kl}\, de_{kl}, \qquad h = \frac{4}{9}(3G + H_p)\sigma_e^2$$

where $H_p = d\sigma_e / d\varepsilon_p$.

Prob. 6.2 Show that for the von Mises loading surface with the isotropic hardening rule, the plastic strain increment can be expressed in term of stress increment as

$$d\varepsilon_{ij}^{p} = \frac{1}{h}\, s_{ij}\, \frac{\partial J_2}{\partial \sigma_{kl}}\, d\sigma_{kl} = \frac{1}{h}\, s_{ij}\, dJ_2, \qquad h = \frac{4}{9}\, H_p \sigma_e^2$$

Prob. 6.3 Show that for the von Mises loading surface with the isotropic hardening rule, the incremental stress-strain relationship can be expressed in the following form

$$d\sigma_{ij} = [\, C_{ijkl} - \frac{4\,G^2}{h}\, s_{ij} s_{kl}\,]\, d\varepsilon_{kl}, \qquad h = \frac{4}{9}\,(3\,G + H_p)\,\sigma_e^2$$

Prob. 6.4 Derive the effective stress and effective plastic strain expressions for the Mohr-Coulomb material. Do not consider singularity on the Mohr-Coulomb surface, and assume $\sigma_1 \geq \sigma_2 \geq \sigma_3$.

Solution: For the Mohr-Coulomb criterion with $\sigma_1 > \sigma_2 > \sigma_3$, Equation (6.23) becomes

$$m\sigma_1 - \sigma_3 = C\, \sigma_e^n$$

For a uniaxial tension test, we note

$$\sigma_1 = \sigma_e\,, \qquad \sigma_3 = 0$$

and this leads to

$$m\sigma_e = C\, \sigma_e^n$$

from which we obtain the constants

$$C = m\,, \qquad n = 1$$

Thus, the effective stress for the Mohr-Coulomb criterion has the form

$$\sigma_e = \frac{m\sigma_1 - \sigma_3}{m}$$

Using Eq. (6.28),

$$dW_p = \frac{\sqrt{d\varepsilon_{ij}^{p}\, d\varepsilon_{ij}^{p}}\; nF}{\sqrt{\dfrac{\partial F}{\partial \sigma_{mn}}\, \dfrac{\partial F}{\partial \sigma_{mn}}}} = \sigma_e d\varepsilon_p$$

with the following relationships for the Mohr-Coulomb criterion,

$$n = 1, \qquad \frac{\partial F}{\partial \sigma_{mn}} \frac{\partial F}{\partial \sigma_{mn}} = m^2 + 1$$

we obtain

$$\frac{\sqrt{d\varepsilon_{ij}^p \, d\varepsilon_{ij}^p}\ (1)\,(m\sigma_1 - \sigma_3)}{\sqrt{m^2+1}} = \sigma_e d\varepsilon_p = \frac{m\sigma_1 - \sigma_3}{m}\, d\varepsilon_p$$

From which the effective plastic strain is defined as

$$d\varepsilon_p = \frac{m}{\sqrt{m^2+1}} \sqrt{d\varepsilon_{ij}^p \, d\varepsilon_{ij}^p}$$

Prob. 6.5 The Tresca and Rankine criterion are special cases of the Mohr-Coulomb criterion. Derive the effective stress and effective plastic strain expressions for Tresca and Rankine functions using the results obtained in Prob. 6.4.

Answer:

Tresca: $\sigma_e = \sigma_1 - \sigma_3, \quad d\varepsilon_p = \sqrt{\frac{1}{2} d\varepsilon_{ij}^p d\varepsilon_{ij}^p}$

Rankine: $\sigma_e = \sigma_1, \quad d\varepsilon_p = \sqrt{d\varepsilon_{ij}^p \, d\varepsilon_{ij}^p}$

Prob. 6.6 Show that for the Drucker-Prager criterion, the effective stress and effective plastic strain expressions are

$$\sigma_e = \frac{\sqrt{3}\,\alpha I_1 + \sqrt{3 J_2}}{1 + \sqrt{3}\,\alpha}$$

$$d\varepsilon_p = \frac{\alpha + 1/\sqrt{3}}{\sqrt{3\alpha^2 + \frac{1}{2}}} \sqrt{d\varepsilon_{ij}^p \, d\varepsilon_{ij}^p}$$

Prob. 6.7 The stress-strain relationship for a material in simple tension has the form

$$\varepsilon = \varepsilon^e + \varepsilon^p = \frac{\sigma}{E}\left[1 + \left(\frac{\sigma}{\sigma_0}\right)^m\right]$$

where σ_0 is the initial yield stress in tension, and m is a material constant. Using the von Mises isotropic surface, derive the expression for $d\tau / d\gamma$ in a pure shear stress state.

Solution: From Prob. 6.2, we have

$$d\varepsilon_{ij}^p = \frac{1}{h}\, s_{ij}\, dJ_2$$

where

$$\frac{1}{h} = \frac{9}{4}\,\frac{1}{H_p \sigma_e^2}, \qquad H_p = \frac{d\sigma_e}{d\varepsilon_p}$$

$$\sigma_e^2 = 3\,J_2, \quad d\varepsilon_p = \sqrt{\frac{2}{3}\, d\varepsilon_{ij}^p\, d\varepsilon_{ij}^p}$$

Using the uniaxial stress-strain curve, we obtain

$$H_p = \frac{E}{m+1}\left(\frac{\sigma_0}{\sigma_e}\right)^m = \frac{E}{m+1}\left(\frac{\sigma_0}{\sqrt{3J_2}}\right)^m$$

and

$$\frac{1}{h} = \frac{3}{4}\,\frac{m+1}{E\,J_2}\left(\frac{\sqrt{3J_2}}{\sigma_0}\right)^m$$

For a pure shear stress state, we have $\tau_{xy} = s_{xy} = \tau$, $J_2 = \tau^2$, $dJ_2 = 2\tau\, d\tau$ and obtain the relationship

$$d\varepsilon_{xy}^p = \frac{3}{4}\,\frac{m+1}{E\,\tau^2}\left(\frac{\sqrt{3}\,\tau}{\sigma_0}\right)^m (\tau)\,(2\tau)\, d\tau = \frac{3(m+1)}{2E}\left(\frac{\sqrt{3}\,\tau}{\sigma_0}\right)^m d\tau$$

and

$$d\gamma = d\gamma^e + d\gamma^p = \frac{d\tau}{G} + \frac{3(m+1)}{E}\left(\frac{\sqrt{3}\,\tau}{\sigma_0}\right)^m d\tau$$

or

$$\frac{d\tau}{d\gamma} = \frac{1}{\dfrac{1}{G} + \dfrac{3(m+1)}{E}\left(\dfrac{\sqrt{3}\,\tau}{\sigma_0}\right)^m}$$

Prob. 6.8 A thin-walled circular tube is subjected to a combined axial-tension and twisting-moment loading. The stress-strain relation of the material in simple tension is given by

$$\varepsilon = \begin{cases} \dfrac{\sigma}{E} & \sigma < \sigma_0 \\ \dfrac{\sigma_0}{E} + \dfrac{\sigma - \sigma_0}{E_p} & \sigma \geq \sigma_0 \end{cases}$$

where E is the elastic modulus, and E_p the plastic modulus, σ_0 the initial yield stress. Assume isotropic von Mises model for the material, and denote the axial stress and axial plastic strain by σ and ε^p, and the shear stress and plastic shear strain τ and γ^p, show that the plastic strain increment expressions are given by

$$d\varepsilon^p = \frac{1}{E_p} \frac{\sigma^2 \, d\sigma + 3\tau\sigma \, d\tau}{\sigma^2 + 3\tau^2}$$

$$d\gamma^p = \frac{1}{E_p} \frac{3\sigma\tau \, d\sigma + 9\tau^2 \, d\tau}{\sigma^2 + 3\tau^2}$$

Prob. 6.9 Same thin-walled tube as in Prob. 6.8 with the final stress state being $\sigma = \sigma_0$ and $\tau = \sigma_0 / \sqrt{3}$, find the corresponding state of strain (ε, γ) by the following loading paths (Fig. P6.6):

a. $(\sigma, \tau) = (0, 0) \rightarrow (\sigma_0, 0) \rightarrow (\sigma_0, \sigma_0 / \sqrt{3})$

b. $(\sigma, \tau) = (0, 0) \rightarrow (0, \sigma_0 / \sqrt{3}) \rightarrow (\sigma_0, \sigma_0 / \sqrt{3})$

c. $(\sigma, \tau) = (0, 0) \rightarrow (\sigma_0, \sigma_0 / \sqrt{3})$

Solution: The expressions for the plastic strain increments $(d\varepsilon^p, d\gamma^p)$ have been derived in Prob. 6.8 as

$$d\varepsilon^p = d\varepsilon_x^p = \frac{1}{E_p} \frac{\sigma^2}{\sigma^2 + 3\tau^2} d\sigma + \frac{3}{E_p} \frac{\tau\sigma}{\sigma^2 + 3\tau^2} d\tau$$

$$d\gamma^p = 2 \, d\varepsilon_{xy}^p = \frac{3}{E_p} \frac{\sigma\tau}{\sigma^2 + 3\tau^2} d\sigma + \frac{1}{E_p} \frac{9\tau^2}{\sigma^2 + 3\tau^2} d\tau$$

The corresponding elastic strains at the final stress state, $(\sigma_0, \sigma_0 / \sqrt{3})$ are

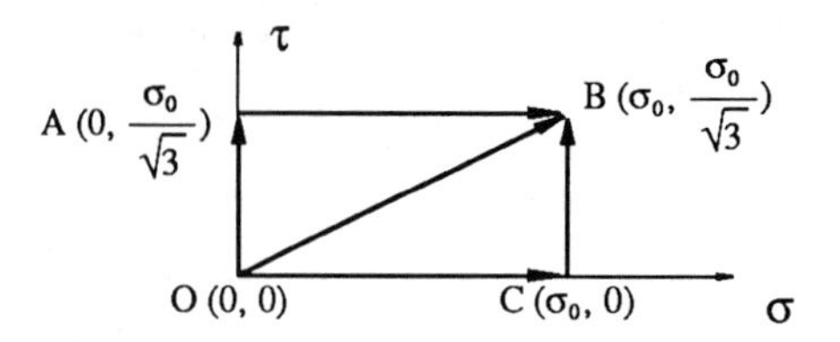

Figure P6.6.

$$\varepsilon^e = \frac{\sigma_0}{E}, \; \gamma^e = \frac{\sigma_0}{\sqrt{3}\,G}$$

Case (a): Path OCB

For $\sigma < \sigma_0$, the tube remains in an elastic state. At State C, $\sigma = \sigma_0$, $\tau = 0$, $\varepsilon^p = \gamma^p = 0$. From C to B, we have $d\sigma = 0$ and $d\tau \neq 0$, thus, we obtain

$$E_p \, d\gamma^p = \frac{9\tau^2}{\sigma_0^2 + 3\tau^2} \, d\tau$$

$$E_p \, \gamma^p = \int_0^{\sigma_0/\sqrt{3}} \frac{9\tau^2}{3\tau^2 + \sigma_0^2} \, d\tau = \sqrt{3} \, \sigma_0 \, (1 - \frac{\pi}{4})$$

and

$$E_p \, d\varepsilon^p = 3\sigma_0 \frac{\tau}{3\tau^2 + \sigma_0^2} \, d\tau$$

$$E_p \, \varepsilon^p = \int_0^{\sigma_0/\sqrt{3}} 3\sigma_0 \frac{\tau}{3\tau^2 + \sigma_0^2} \, d\tau = \sigma_0 \, ln \sqrt{2}$$

At the final state, we have

$$\varepsilon = \varepsilon^e + \varepsilon^p = \sigma_0 \, [\frac{1}{E} + \frac{1}{E_p} \, ln \sqrt{2}]$$

$$\gamma = \gamma^e + \gamma^p = \frac{\sigma_0}{\sqrt{3}} \, [\frac{1}{G} + \frac{3}{E_p} \, (1 - \frac{\pi}{4})]$$

Case (b): Path OAB

$\tau < \sigma_0/\sqrt{3}$, the tube remains in an elastic state, and at State A, $\tau = \sigma_0 / \sqrt{3}$, $\sigma = 0$, $\varepsilon^p = \gamma^p = 0$. From A to B, we have $d\tau = 0$, $d\sigma \neq 0$, and thus obtain,

$$E_p \, d\varepsilon^p = \frac{\sigma^2}{\sigma^2 + 3\tau^2} \, d\sigma = \frac{\sigma^2}{\sigma^2 + \sigma_0^2} \, d\sigma$$

$$E_p \varepsilon^p = \int_0^{\sigma_0} \frac{\sigma^2}{\sigma^2 + \sigma_0^2} \, d\sigma = \sigma_0 \left(1 - \frac{\pi}{4}\right)$$

and

$$E_p \, d\gamma^p = \frac{3\sigma\tau}{\sigma^2 + 3\tau^2} \, d\sigma = \frac{\sqrt{3}\,\sigma_0\,\sigma}{\sigma^2 + \sigma_0^2} \, d\sigma$$

$$E_p \, \gamma^p = \int_0^{\sigma_0} \sqrt{3}\,\sigma_0 \frac{\sigma}{\sigma^2 + \sigma_0^2} \, d\sigma = \sqrt{3}\,\sigma_0 \, ln \sqrt{2}$$

At the final state, we have

$$\varepsilon = \varepsilon^e + \varepsilon^p = \sigma_0 \left[\frac{1}{E} + \frac{1}{E_p}\left(1 - \frac{\pi}{4}\right)\right]$$

$$\gamma = \gamma^e + \gamma^p = \frac{\sigma_0}{\sqrt{3}} \left[\frac{1}{G} + \frac{3}{E_p} \, ln \sqrt{2}\right]$$

Case (c): Path OB

Along the loading path OB, we have $\sigma/\tau = \sqrt{3}$. The initial yielding occurs at $\sigma = \sigma_0/\sqrt{2}$, $\tau = \sigma_0/\sqrt{6}$. Beyond the initial yielding, we have $d\sigma = \sqrt{3}\, d\tau$, thus obtain

$$E_p \, d\varepsilon^p = \frac{\sigma^2}{\sigma^2 + 3\tau^2} \, d\sigma + \frac{3\tau\sigma}{\sigma^2 + 3\tau^2} \, d\tau = d\sigma$$

$$E_p \varepsilon^p = \int_{\sigma_0/\sqrt{2}}^{\sigma_0} d\sigma = \sigma_0 \left(1 - \frac{1}{\sqrt{2}}\right)$$

and

$$E_p \, d\gamma^p = \frac{3\sigma\tau}{\sigma^2 + 3\tau^2} \, d\sigma + \frac{9\tau^2}{\sigma^2 + 3\tau^2} \, d\tau = \sqrt{3} \, d\sigma$$

$$E_p \, \gamma^p = \int_{\sigma_0/\sqrt{2}}^{\sigma_0} \sqrt{3} \, d\sigma = \sqrt{3}\,\sigma_0 \left(1 - \frac{1}{\sqrt{2}}\right)$$

Thus, at the final state, we have

$$\varepsilon = \varepsilon^e + \varepsilon^p = \sigma_0 \left[\frac{1}{E} + \frac{1}{E_p}\left(1 - \frac{1}{\sqrt{2}}\right)\right]$$

$$\gamma = \gamma^e + \gamma^p = \frac{\sigma_0}{\sqrt{3}} \left[\frac{1}{G} + \frac{3}{E_p}\left(1 - \frac{1}{\sqrt{2}}\right)\right]$$

Prob. 6.10 A long thin-walled steel tube with diameter D and wall thickness t subjected to an internal pressure p_1 and an external pressure p_2 is shown in Fig. P6.7. The ends of the tube are closed. The external pressure is assumed not to affect the axial stress component of the tube. Assume the von Mises loading function with isotropic hardening rule for the material. Use the effective stress - effective plastic strain relationship of the form

$$\varepsilon_p = a\,\sigma_e^3$$

where a is a constant. Determine the plastic strains $(\varepsilon_a^p, \varepsilon_c^p)$ at the end of the following three loading paths:

a. $(p_1, p_2) = (0, 0) \rightarrow (P_1, RP_1)$
b. $(p_1, p_2) = (0, 0) \rightarrow (0, RP_1) \rightarrow (P_1, RP_1)$
c. $(p_1, p_2) = (0, 0) \rightarrow (P_1, 0) \rightarrow (P_1, RP_1)$

where ε_a^p and ε_c^p are the axial and circumferential plastic strains, respectively, and P_1 and R are constants, R = 3/2. Sketch the yield surfaces and the loading paths in the (σ_a, σ_c) space, and explain the results so obtained in the three loading cases.

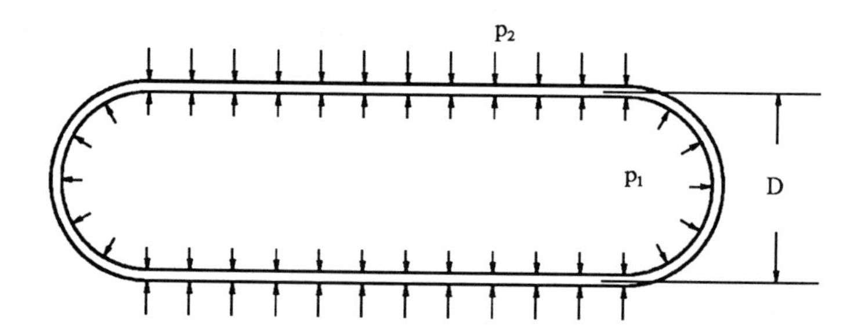

Figure P6.7.

Solution: The thin-walled tube is in a biaxial stress state, and the corresponding axial and circumferential components, σ_a and σ_c, are principal stresses with

$$\sigma_a = \frac{D}{4t}\,p_1, \quad \sigma_c = \frac{D}{2t}\,(p_1 - p_2)$$

and we have

$$J_2 = \frac{1}{3}\,(\sigma_a^2 + \sigma_c^2 - \sigma_a\sigma_c), \quad \sigma_e^2 = 3J_2 = \sigma_a^2 + \sigma_c^2 - \sigma_a\sigma_c$$

$$dJ_2 = \frac{1}{3}\,(2\sigma_a - \sigma_c)\,d\sigma_a + \frac{1}{3}\,(2\sigma_c - \sigma_a)\,d\sigma_c$$

From Prob. 6.2, we have the plastic strain increments

$$d\varepsilon_{ij}^p = \frac{1}{h}\, s_{ij}\, dJ_2$$

where

$$\frac{1}{h} = \frac{9}{4}\,\frac{1}{H_p\sigma_e^2}, \qquad H_p = \frac{d\sigma_e}{d\varepsilon_p}$$

and for this problem, we have

$$H_p = \frac{1}{3a\sigma_e^2}, \qquad \frac{1}{h} = \frac{27a}{4}$$

therefore, we obtain

$$d\varepsilon_{ij}^p = \frac{9a}{4}\,[(2\sigma_a - \sigma_c)\, d\sigma_a + (2\sigma_c - \sigma_a)\, d\sigma_c]\, s_{ij}$$

Case (a): Proportional Loading Path, $(0, 0) \to (P_1, RP_1)$.

For this case, we have $p_2 = RP_1 = \frac{3}{2}\, p_1$, and

$$\sigma_a = \frac{D}{4t}\, p_1, \quad \sigma_c = -\frac{D}{4t}\, p_1 = -\sigma_a, \quad d\sigma_c = -\, d\sigma_a$$

and obtain $s_a = \sigma_a$, $s_c = \sigma_c$, and

$$d\varepsilon_{ij}^p = \frac{9a}{4}\, 6\sigma_a\, d\sigma_a\, s_{ij} = \frac{27a}{2}\, \sigma_a\, d\sigma_a\, s_{ij}$$

In the component form, we have

$$d\varepsilon_a^p = \frac{27a}{2}\, \sigma_a^2\, d\sigma_a = \frac{27a}{2}\, (\frac{D}{4t})^3\, p_1^2\, dp_1$$

$$d\varepsilon_c^p = -\frac{27a}{2}\, \sigma_a^2\, d\sigma_a = -\frac{27a}{2}\, (\frac{D}{4t})^3\, p_1^2\, dp_1$$

Integrate the above expressions and obtain the plastic strains at the end of this loading path as

$$\varepsilon_a^p = \frac{27a}{2}\, (\frac{D}{4t})^3 \int_0^{P_1} p_1^2\, dp_1 = \frac{9a}{2}\, (\frac{DP_1}{4t})^3$$

$$\varepsilon_c^p = -\frac{27a}{2}\, (\frac{D}{4t})^3 \int_0^{P_1} p_1^2\, dp_1 = -\frac{9a}{2}\, (\frac{DP_1}{4t})^3$$

Case (b): Non-Proportional Loading Path, $(0,0) \to (0, RP_1) \to (P_1, RP_1)$

First increasing p_2 from 0 to RP_1, we have

$$\sigma_a = 0,\ \sigma_c = -\frac{D}{2t}\,p_2,\ d\sigma_a = 0,\ d\sigma_c \neq 0.$$

and we have $s_a = -\frac{1}{3}\,\sigma_c,\ s_c = \frac{2}{3}\,\sigma_c$ and

$$d\varepsilon_{ij}^{p} = \frac{9a}{4}\,2\sigma_c\,d\sigma_c\,s_{ij} = \frac{9a}{2}\,\sigma_c\,d\sigma_c\,s_{ij}$$

In the component form, we have

$$d\varepsilon_a^p = \frac{9a}{2}\,\sigma_c\,d\sigma_c(-\frac{1}{3}\,\sigma_c) = -\frac{3a}{2}\,\sigma_c^2\,d\sigma_c = \frac{3a}{2}\,(\frac{D}{2t})^3\,p_2^2\,dp_2$$

$$d\varepsilon_c^p = \frac{9a}{2}\,\sigma_c\,d\sigma_c\,(\frac{2}{3}\,\sigma_c) = 3a\,\sigma_c^2\,d\sigma_c = -\,3a\,(\frac{D}{2t})^3\,p_2^2\,dp_2$$

At $p_1 = 0$, $p_2 = RP_1$, we obtain

$$\varepsilon_a^p = -\frac{3a}{2}\,(\frac{D}{2t})^3\int_0^{RP_1} p_2^2\,dp_2 = \frac{a}{2}\,(\frac{DRP_1}{2t})^3 = \frac{a}{2}\,(\frac{3DP_1}{4t})^3$$

$$\varepsilon_c^p = -\,3a\,(\frac{D}{2t})^3\int_0^{RP_1} p_2^2\,dp_2 = -\,a\,(\frac{DRP_1}{2t})^3 = -\,a\,(\frac{3DP_1}{4t})^3$$

and

$$J_2 = \frac{1}{3}\,\sigma_c^2 = \frac{1}{3}\,(\frac{D}{2t})^2\,R^2\,P_1^2 = \frac{3}{4}\,(\frac{D}{2t})^2\,P_1^2$$

In this path, increasing p_1 from 0 to P_1 and $p_2 = RP_1$, we have

$$\sigma_a = \frac{D}{4t}\,p_1,\ \sigma_c = \frac{D}{2t}\,p_1 - \frac{D}{2t}\,RP_1 = 2\sigma_a - \frac{D}{2t}\,RP_1$$

Since the value of σ_c has been reduced in the present path, we need check the loading criterion first. Let J_2 at the beginning of this path equal to the current J_2, we obtain

$$\frac{1}{3}\,[\sigma_a^2 + (2\sigma_a - \frac{D}{2t}\,RP_1)^2 - \sigma_a\,(2\sigma_a - \frac{D}{2t}\,RP_1)] = \frac{3}{4}\,(\frac{D}{2t})^2\,P_1^2$$

or

$$\sigma_a^2 - \sigma_a\,\frac{D}{2t}\,RP_1 = 0$$

Solving the above equation for σ_a, we find

$$\sigma_a = 0,\ \text{or}\ \sigma_a = \frac{D}{2t}\,RP_1 = \frac{3D}{4}\,P_1 > \frac{D}{4t}\,P_1$$

This implies that for an increasing p_1 from 0 to P_1, the tube is either in an unloading state or in an elastic loading state, and no new plastic deformations will be produced. Therefore, at the end of this path, we have

$$\varepsilon_a^p = \frac{27a}{2}\left(\frac{DP_1}{4t}\right)^3 \qquad \varepsilon_c^p = -27a\left(\frac{DP_1}{4t}\right)^3$$

Case (c): Non-Proportional Loading Path, $(0, 0) \to (P_1, 0) \to (P_1, RP_1)$

In this path, increasing p_1 from 0 to P_1, we have

$$\sigma_a = \frac{D}{4t}p_1,\ \ \sigma_c = \frac{D}{2t}p_1 = 2\sigma_a,\ \ d\sigma_c = 2\,d\sigma_a$$

and obtain $s_a = 0$, $s_c = \sigma_a$, and

$$d\varepsilon_{ij}^p = \frac{9a}{4}6\sigma_a\,d\sigma_a\,s_{ij} = \frac{27a}{2}\sigma_a\,d\sigma_a\,s_{ij}$$

In the component form, we have

$$d\varepsilon_a^p = 0$$

$$d\varepsilon_c^p = \frac{27a}{2}\sigma_a^2\,d\sigma_a = \frac{27a}{2}\left(\frac{D}{4t}\right)^3 p_1^2\,dp_1$$

At $p_1 = P_1$, $p_2 = 0$, we have

$$\varepsilon_a^p = 0$$

$$\varepsilon_c^p = \frac{27a}{2}\left(\frac{D}{4t}\right)^3\int_0^{P_1} p_1^2\,dp_1 = \frac{9a}{2}\left(\frac{DP_1}{4t}\right)^3$$

and

$$J_2 = \sigma_a^2 = \left(\frac{D}{4t}\right)^2 P_1^2$$

In this path, increasing p_2 from 0 to RP_1, and $p_1 = P_1$, we have

$$\sigma_a = \frac{D}{4t}P_1,\ \ \sigma_c = \frac{D}{2t}P_1 - \frac{D}{2t}p_2 = 2\sigma_a - \frac{D}{2t}p_2$$

Since σ_c has been reduced in the present path, we need check the loading criterion first. Let J_2 at the beginning of this path equal to the current J_2, we have

$$\frac{1}{3}\left[\sigma_a^2 + \left(2\sigma_a - \frac{D}{2t}p_2\right)^2 - \sigma_a\left(2\sigma_a - \frac{D}{2t}p_2\right)\right] = \sigma_a^2$$

or

$$\left(\frac{Dp_2}{2t}\right)^2 - 3\sigma_a\left(\frac{Dp_2}{2t}\right) = 0$$

Solving for p_2, we obtain

$$p_2 = 0,\ \ p_2 = \frac{3}{2}P_1 = P_2$$

This implies that in this path, increasing p_2 from 0 to P_2, the tube is either in an unloading state or in an elastic loading state, and no new plastic deformations will be produced. Therefore, at the end of this path, we have

$$\varepsilon_a^p = 0, \quad \varepsilon_c^p = \frac{9a}{2}\left(\frac{DP_1}{4t}\right)^3$$

The yield surfaces and loading paths are plotted in Fig. S6.7. The normal direction to the surfaces can be expressed as

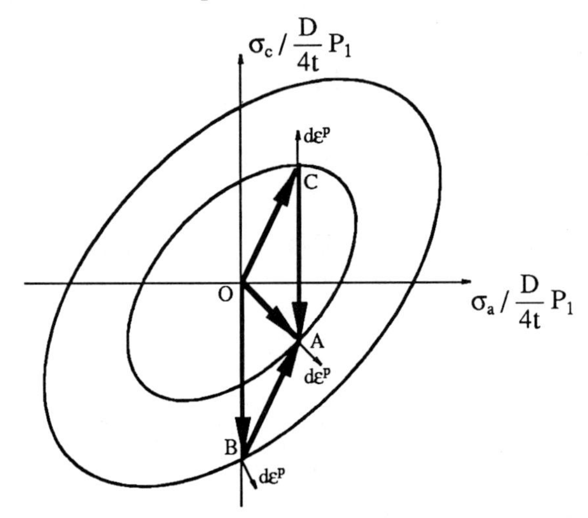

Figure S6.7.

$$\vec{n} = \left(\frac{\partial J_2}{\partial \sigma_a}, \frac{\partial J_2}{\partial \sigma_c}\right) = \left(\frac{2\sigma_a - \sigma_c}{3}, \frac{2\sigma_c - \sigma_a}{3}\right)$$

Path OA is a proportional loading path,

$$\sigma_c = -\sigma_a$$

or

$$\vec{n} = (\sigma_a, -\sigma_a)$$

and therefore, along this path, we always have $d\varepsilon_c^p = -d\varepsilon_a^p$ as in Case (a).

For Path OB, the loading $(p_1, p_2) = (0, 0) \rightarrow (0, RP_1)$, we have $\sigma_a = 0$, $\sigma_c < 0$, and

$$\vec{n} = \left(-\frac{\sigma_c}{3}, \frac{2}{3}\sigma_c\right)$$

therefore along this path, we have $d\varepsilon_a^p > 0$, $d\varepsilon_c^p = -2\, d\varepsilon_a^p$, as in Case (b).

For Path OC, the loading path $(p_1, p_2) = (0, 0) \rightarrow (P_1, 0)$, we have $\sigma_c = 2\sigma_a$, and

$$\vec{n} = (0,\ \sigma_a)$$

therefore along this path, we have $d\varepsilon_a^p = 0$, $d\varepsilon_c^p > 0$, as in Case (c).

Prob. 6.11 Assume the von Mises loading function with isotropic hardening rule. Use the effective stress - effective plastic strain relationship

$$\varepsilon_p = a\,\sigma_e^3$$

where a is a constant. An element of the material is subjected to the following three loading paths in the principal stress space (σ_1, σ_2, σ_3):

a. $(\sigma_1, \sigma_2, \sigma_3) = O(0, 0, 0) \rightarrow A(\sigma_0, \sigma_0, 0) \rightarrow C(\sigma_0, \sigma_0, 3\sigma_0)$
b. $(\sigma_1, \sigma_2, \sigma_3) = O(0, 0, 0) \rightarrow B(0, 0, 3\sigma_0) \rightarrow C(\sigma_0, \sigma_0, 3\sigma_0)$
c. $(\sigma_1, \sigma_2, \sigma_3) = O(0, 0, 0) \rightarrow C(\sigma_0, \sigma_0, 3\sigma_0)$

where σ_0 is a constant. Determine the corresponding plastic strain at the end of each loading path. Noting that in all three loading paths, we have $\sigma_1 = \sigma_2$.

Answer:

$$\text{(a)}\ \ \varepsilon_1^p = \varepsilon_2^p = -3\,a\,\sigma_0^3, \quad \varepsilon_3^p = 6\,a\,\sigma_0^3$$

$$\text{(b)}\ \ \varepsilon_1^p = \varepsilon_2^p = -\frac{27}{2}\,\sigma_0^3, \quad \varepsilon_3^p = 27\,a\,\sigma_0^3$$

$$\text{(c)}\ \ \varepsilon_1^p = \varepsilon_2^p = -4\,a\,\sigma_0^3, \quad \varepsilon_3^p = 8\,a\,\sigma_0^3$$

6.9 Problems Using Deformational Theory

Prob. 6.12 The stress-strain relation of a J_2-material in a pure shear test is approximated by

$$\gamma = \begin{cases} \dfrac{\tau}{G} & \tau \le \tau_0 \\[2ex] \dfrac{\tau}{G} + \dfrac{\tau - \tau_0}{m} & \tau > \tau_0 \end{cases}$$

where τ_0 is the initial yield stress in shear. Using the J_2-deformational theory, write the complete stress-strain relationships of the material

a. In simple tension state, $\sigma_x = \sigma$, $\varepsilon_x = \varepsilon$, all other stresses are zero; and
b. In equal biaxial tension-compression state, $\sigma_x = \sigma$, $\sigma_y = -\sigma$, $\varepsilon_x = \varepsilon$, $\varepsilon_y = -\varepsilon$, all other stresses are zero.

Solution: For the pure shear stress state, we have $\tau_{xy} = \tau$, $J_2 = \tau^2$, and all other stresses are zero. Using the deformational theory, we have

$$\gamma^p = \gamma^p_{xy} = \frac{3\,\varepsilon_p}{\sigma_e}\,\tau_{xy} = \frac{3\varepsilon_p}{\sigma_e}\,\tau$$

and from the test curve, we have

$$\gamma^p = \frac{\tau - \tau_0}{m}$$

Comparing the two expressions, we obtain

$$\frac{\varepsilon_p}{\sigma_e} = \frac{\tau - \tau_0}{3m\tau} = \frac{\sqrt{J_2} - \tau_0}{3m\,\sqrt{J_2}}$$

Case (a): Simple Tension

For a simple tension stress state, we have $J_2 = \sigma^2/3$, and

$$\varepsilon^p = \varepsilon^p_x = \frac{\sigma/\sqrt{3} - \tau_0}{3\,m\,\sigma/\sqrt{3}}\,[\sigma_x - \frac{1}{2}\,(\sigma_y + \sigma_z)] = \frac{\sigma/\sqrt{3} - \tau_0}{3\,m\,\sigma/\sqrt{3}}\,\sigma$$

$$= \frac{\sigma - \sqrt{3}\,\tau_0}{3m}$$

The complete stress-strain relation is

$$\varepsilon = \begin{cases} \dfrac{\sigma}{E} & \sigma \le \sqrt{3}\,\tau_0 \\[2ex] \dfrac{\sigma}{E} + \dfrac{\sigma - \sqrt{3}\,\tau_y}{3\,m} & \sigma > \sqrt{3}\,\tau_y \end{cases}$$

Case (b): Equal Biaxial Tension-Compression

For this stress state, we have $J_2 = \sigma^2$, and

$$\varepsilon^p = \varepsilon^p_x = \frac{\sigma - \tau_0}{3m\sigma}\,[\sigma_x - \frac{1}{2}\,(\sigma_y + \sigma_z)] = \frac{\sigma - \tau_0}{3m\sigma}\,\frac{3}{2}\,\sigma = \frac{\sigma - \tau_0}{2m}$$

The complete stress-strain relation is

$$\varepsilon = \begin{cases} \dfrac{\sigma(1+\nu)}{E} & \sigma \le \tau_0 \\ \dfrac{\sigma(1+\nu)}{E} + \dfrac{\sigma-\tau_0}{2m} & \sigma > \tau_0 \end{cases}$$

Prob. 6.13 Solve the same thin-walled tube problem as in Prob. 6.9 using the J_2-deformational theory.

Solution: Using Eq. (6.8) and noting that the non-zero stress components are σ_x and τ_{xy}, we have

$$\varepsilon^p = \varepsilon_x^p = \frac{\varepsilon_p}{\sigma_e}\sigma_x = \frac{\varepsilon_p}{\sigma_e}\sigma$$

$$\gamma^p = \gamma_{xy}^p = \frac{3\varepsilon_p}{\sigma_e}\tau_{xy} = \frac{3\varepsilon_p}{\sigma_e}\tau$$

At the final state, we have $\sigma_e = \sqrt{2}\,\sigma_0$, and

$$\frac{\varepsilon_p}{\sigma_e} = \frac{\sigma_e-\sigma_0}{E_p\sigma_e} = \frac{1}{E_p}(1-\frac{1}{\sqrt{2}})$$

At the final state B, we have

$$\varepsilon = \varepsilon^e + \varepsilon^p = \frac{\sigma_0}{E} + \frac{1}{E_p}(1-\frac{1}{\sqrt{2}})\sigma_0$$

$$= \sigma_0 [\frac{1}{E} + \frac{1}{E_p}(1-\frac{1}{\sqrt{2}})]$$

$$\gamma = \gamma^e + \gamma^p = \frac{\sigma_0}{\sqrt{3}\,G} + \frac{3}{E_p}(1-\frac{1}{\sqrt{2}})\frac{\sigma_0}{\sqrt{3}}$$

$$= \frac{\sigma_0}{\sqrt{3}}[\frac{1}{G} + \frac{3}{E_p}(1-\frac{1}{\sqrt{2}})]$$

It can be seen that for the proportional loading case, the deformational theory leads to the same results as in Case (c), Prob. 6.9 as it should.

Prob. 6.14 Solve the same thin-walled tube problem as in Prob. 6.10, using the J_2-deformational theory.

Answer:

$$\varepsilon_a^p = \frac{9a}{2}\left(\frac{DP_1}{4t}\right)^3, \qquad \varepsilon_c^p = -\frac{9a}{2}\left(\frac{DP_1}{4t}\right)^3$$

Prob. 6.15 Solve the same problem as in Prob. 6.11, using the J_2-deformational theory.

Answer:

$$\varepsilon_1^p = \varepsilon_2^p = -4a\sigma_0^3, \quad \varepsilon_3^p = 8a\sigma_0^3$$

Index